IVORYBILL HUNTERS

IVORYBILL HUNTERS

The Search for Proof in a Flooded Wilderness

Geoffrey E. Hill

OXFORD
UNIVERSITY PRESS

2007

Oxford University Press, Inc., publishes works that further
Oxford University's objective of excellence
in research, scholarship, and education.

Oxford   New York
Auckland   Cape Town   Dar es Salaam   Hong Kong   Karachi
Kuala Lumpur   Madrid   Melbourne   Mexico City   Nairobi
New Delhi   Shanghai   Taipei   Toronto

With offices in
Argentina   Austria   Brazil   Chile   Czech Republic   France   Greece
Guatemala   Hungary   Italy   Japan   Poland   Portugal   Singapore
South Korea   Switzerland   Thailand   Turkey   Ukraine   Vietnam

Copyright © 2007 by Oxford University Press, Inc.

Published by Oxford University Press, Inc.
198 Madison Avenue, New York, New York 10016

www.oup.com

Oxford is a registered trademark of Oxford University Press

Library of Congress Cataloging-in-Publication Data
Hill, Geoffrey E. (Geoffrey Edward)
Ivorybill hunters : the search for proof in a flooded wilderness / Geoffrey E. Hill.
  p. cm.
ISBN: 978-0-19-532346-7
1. Ivory-billed woodpecker—Research—Choctawhatchee
River Watershed (Ala. and Fla.)   I. Title.
QL696.P56H55 2007
598.7'2—dc22   2006035191

9 8 7 6 5 4 3 2 1

Printed in the United States of America
on acid-free, recycled paper

Dedicated to Savannah and Trevor
for putting up with a dad
who was always in the swamp.

And to Wendy for sharing the ups
and downs of one crazy year.

The Ivory-billed Woodpecker holds a special place in the lore of North American birds. It is a bird with pizzazz—huge with brilliant black and white feathers and a comical nasal call. It is a bird of mystique—dwelling in the most remote and inhospitable southern swamps. And it is a bird that disappeared in the twentieth century right before the eyes of an America with a growing environmental conscience. A handful of ornithologists and bird enthusiasts refused to accept that the great woodpecker was extinct. They pursued the shadow of the bird into the farthest reaches of southern swamps. News of a few Ivory-billed Woodpeckers in the rapidly dwindling pine forests of Cuba in the late twentieth century rekindled hope that the species could yet be saved, but it was folly to believe that the ivorybill could persist on a crowded Caribbean Island when it had already been squeezed off a continent. The bird again faded to nil, this time, it seemed, for good. But hope that the species might still linger, unrealistic as it was, kept a few searchers in the swamps.

Perhaps the primary keepers of the flame of hope that ivorybills might have survived into the new millennium were John Fitzpatrick and the Cornell Laboratory of Ornithology. Beginning with a widely publicized search of Louisiana's Pearl River in 2002, Fitzpatrick and his colleagues at the Laboratory of Ornithology were the catalysts for a resurgence of interest in the Ivory-billed Woodpecker at the dawn of the twenty-first century. Then, in the spring of 2005, Fitzpatrick announced that a team of searchers had amassed proof that at least one Ivory-billed Woodpecker still dwelled in the vast forested wetlands of eastern Arkansas. The news was scintillating and captured the imaginations of not just birdwatchers and nature enthusiasts but seemingly everyone—from politicians to dentists. It's not every day that a bird sighting makes the front page of every major newspaper in the country and is a top story on the evening news. Unfortunately, the supposedly irrefutable evidence for an ivorybill in Arkansas was refuted by some of the most respected woodpecker and bird identification experts in North America. In the face of building skepticism, the claim of an Ivory-billed Woodpecker in Arkansas faded from focus rather like the image on the video that purported to show an ivorybill flying near the Cache River. It was almost too cruel to have the bird taken away once again. But regardless of whether one accepted the proof of an Arkansas ivorybill, the announcement by Fitzpatrick spurred more people to take up the search, penetrating deeper into the swamps and searching areas that had previously been overlooked.

This book is the tale of a search for ivorybills far removed from the highly publicized searches in Arkansas. I focus primarily on the search and the searchers rather than on the woodpecker. After all, the search for and rediscovery of the Ivory-billed Woodpecker is a human tale. The ivorybills of the Choctawhatchee River wetlands in the panhandle of northern Florida eat their grubs, cut their cavities, and raise their young as they have always done. That a few humans took note of them in May 2005 certainly is of no consequence to the birds. Humans have always been a nuisance; human intrusions simply got a little worse after 2005.

The new-millennium searches for ivorybills along the Pearl River in Louisiana and especially along the Cache and White rivers in Arkansas attracted a lot of media attention. Public interest in these searches has been enormous, but so too has been the level of misunderstanding by the birding public regarding the role of amateurs versus professionals in the ivorybill search and the process of peer review and publication in a journal like *Science*. As a professional ornithologist I will add some perspective on what science can and can't contribute to a search for a rare animal such as the Ivory-billed Woodpecker.

My story takes place in one small region of the Deep South far removed from other ivorybill hunts. It has about ten main characters and covers a 12-month period in 2005 and 2006. It is the tale of three ivorybill hunters from Auburn University who found this most elusive species on their second morning searching in a place no expert had ever mentioned as a good place to look. Because I was one of the three discoverers, and I played a central role in both the discovery and the subsequent search for ivorybills, I thought I should write down the events as I saw them unfold. There seems to be an insatiable thirst for ivorybill stories among birders and nature lovers, and I've never been one to pass on a chance to tell a tale.

Throughout this book I repeatedly use the names "Ivory-billed Woodpecker" and "ivorybill." Ivory-billed Woodpecker is the correct common name of *Campephilus principalis* as stated by the American Ornithologists' Union *Check-list of North American Birds,* seventh edition. Following convention of North American ornithologists I capitalize this name when used in full. When I shorten the name to ivorybill, it is no longer capitalized. Some authors hyphenate this shorthand form of the common name of *Campephilus principalis,* but convention in North American ornithology is not to hyphenate such abbreviated names. For instance, we would write redwing as short for Red-winged Blackbird and rubythroat as short for Ruby-throated Hummingbird. The justification for not hyphenating such names is explained in detail in a 1937 article in *The Auk* by William Cheesman and Paul Oehser.

This book is a story, not a technical report. I use no citations or even footnotes. I present no statistical comparisons of data sets or quantitative analyses of the sounds that we recorded. Readers who want technical details, statistical analyses, and references supporting stated facts and assertions should refer to the journal article summarizing the evidence for the presence of ivorybills that Dan Mennill, Brian Rolek, Kyle Swiston, Tyler Hicks, and I published on September 26, 2006 in the on-line journal *Avian Conservation and Ecology* (http://www.ace-eco.org). The appendix at the end of this book presents tables summarizing the times and dates of our encounters with ivorybills.

Events in this book are told to my best recollection. My friends and co-workers including Paul Mennill, Brian Rolek, Kyle Swiston, Tyler Hicks, and Sylvie Tremblay read this book and corrected factual errors. In the few instances in which there were differences in recollections of events, you get my account. Special thanks to Dan Mennill, Ken Able, Peter Prescott, Wendy Hood, Chuck Hunter, Mark Bailey, and my mother, Sally Hill, for detailed comments on the entire text, which greatly improved the writing. Jerry Jackson and two anonymous readers reviewed the book and provided many useful suggestions. The opinions expressed in this book are mine.

CONTENTS

IVORYBILL HUNTERS

Gone but Not Forgotten | 1

Sometime in 1995, I received a call transferred by the departmental secretary to my office. I had been at my job as an assistant professor at Auburn University in southeast Alabama for about two years. My Auburn job was the fulfillment of a dream for me. I was not just a faculty member and not merely a biologist—I was officially the school's "avian biologist." From my first day at Auburn I was the primary bird man at the university, and, by default, I was in a sense the state ornithologist. The University of Alabama at Birmingham (UAB), the University of Alabama (UA), and Auburn University are the three major universities in Alabama, and neither UAB nor UA had a true ornithologist on their faculty. Virtually any bird sighting or question that started with a county agent or local college in Alabama would find its way to me.

"Hello, this is Dr. Hill," I answered, using my title since I knew I was getting a "bird call" from the general public. I would usually just answer my phone as "Geoff."

"Are you a bird expert?" a man with a south Alabama accent asked.

"Yes, I'm the ornithologist at Auburn," I answered.

"Well, I saw an Ivory-billed Woodpecker this morning from my truck," the man said in a somewhat defiant voice.

"Really?" I responded, trying not to sound skeptical. "Where did you see it?"

"I live in Geneva County. I saw the bird on a post, then it flew off to the woods," the caller stated matter-of-factly.

3

"What made you think it was an ivorybill?" I asked, wondering if this was going to quickly erode into a sighting of a Pileated Woodpecker.

"It was huge and black and white with a light-colored bill," he answered.

"There's another big woodpecker in Geneva County. Are you sure it wasn't a Pileated Woodpecker?" I asked, probing to see if he knew anything about birds.

"Look Doc. I've been seein' pileateds every day of my life. I hunt all the time and I damn well know what a pileated looks like. They nest in my yard. This weren't no pileated. It was an ivorybill. It had a big white triangle on its back when it perched, and its bill was creamy white."

"Is there any swamp forest around there?"

"You bet. It flew off toward the Pea River. That woods right there along the Pea is 7,000 acres with lots of old timber."

"I believe you, but I don't know what to do about it," I said, voicing my uncertainty. I really did believe he had probably seen an ivorybill, but with no proof, what was I supposed to do? "You know that everyone thinks that bird is extinct."

"I know it, Doc. That's why I'm calling to report it."

"All I can say is carry a video camera and tape recorder with you and try to get pictures or sound. That's the only way people will believe that ivory-bills survive in Alabama."

"Ok. I'll put my camera in my truck, but I don't know if I'll get another chance."

"Thanks for letting me know about the sighting, and good luck. Call back if you see it again or especially if you get a picture of it."

"Thanks for listenin' doc."

I put down the phone, jotted a few notes, and went on with my business. I was too busy with my new faculty position to follow up, and even though I was a professional ornithologist, I didn't know whom to call to officially log the sighting. It was like getting a report of Bigfoot or the Loch Ness Monster. As far as I knew, there was no place to log sightings of extinct birds.

The Ivory-billed Woodpecker was not just any extinct bird. This was no Dusky Seaside Sparrow, a little brown bird now extinct in the United States that was so subtly different from its abundant cousin the Seaside Sparrow that even experts had to look hard to tell the difference. The Ivory-billed Woodpecker was one hell of a bird. With a wingspan of almost three feet

A male Ivory-billed Woodpecker photographed at its nest in the Singer Tract in Louisiana in 1935. (Photograph by Arthur A. Allen, ©Cornell Laboratory of Ornithology.)

A female ivorybill at her nest in the Singer Tract in 1935. Females are identical to males in plumage pattern, except they lack red on the crest. (Photograph by Arthur A. Allen, ©Cornell Laboratory of Ornithology.)

and striking black and white plumage, it had prompted the name Lord God Bird or simply Lord God in many southern communities, where it was well known until the early twentieth century. To the birding novice, ivorybills were superficially similar to Pileated Woodpeckers, another big, mostly black woodpecker with a crest that shared southern swamp forests with ivorybills. Ivorybills are about 10–20% bigger than a pileateds, but it wasn't really size that set them apart. The plumage of the two woodpeckers was certainly different. The pileated had two rather small patches of white toward the center of the upper side of the wings, whereas the ivorybill had a broad band of brilliant white across the back (trailing edge) of its wings. Everyone who had the privilege to witness an Ivory-billed Woodpecker flying through a forest noted this conspicuous trailing band of white. Female ivorybills had an all-black crest, but male ivorybills, like both male and female pileateds, had red on their crests.

Plumage was a great identification aid for birders, but black and white feathers did not earn the ivorybill the title "Lord God Bird." Ivorybills were the Lord God of birds because of their demeanor. To those that observed them, these birds embodied power and nobility. Ivorybills flew straight and powerfully, and they could chop through solid wood almost as fast as a man with an axe. This was a bird that commanded respect.

Ivory-billed Woodpeckers had been relatively common denizens of southern swamp forests before the Civil War. Two hundred years or so of encroachment by Europeans had diminished to some degree the great stands of cypress, tupelo, and oak that grew in low areas and along major rivers throughout the South, but many—perhaps most—of these grand forests stood untouched in the mid-nineteenth century. Following the Civil War, however, the devastation of the cypress swamps began in earnest. The industrial age was in high gear, and the ever-expanding urban centers of the North needed a cheap supply of lumber. The South was on its economic knees, in desperate need of cash and jobs, and in no position to protect its natural resources. Northern companies began to buy up and cut down vast tracts of virgin swamp forest. River systems were cut from their deltas to their headwaters using poor southerners, both black and white, as an inexhaustible source of cheap labor. Even vast and seemingly impenetrable swamps such as the Okefenokee in Georgia and Big Cypress in Florida were crisscrossed with canals and rail lines, and the trees were harvested. This orgy of consumption ran unabated from the mid-nineteenth century to the mid-twentieth century, and it left scarcely an acre of virgin cypress on the entire continent. Greed knows no bounds, and the complete consumption of the vast cypress forests (and simultaneously the long-leaf pine forests in the uplands surrounding the cypress swamps) in less than a century stands as one of the greatest feats of resource gluttony in American history.

With the great swamp forests went the ivorybills. In areas where the forests were converted to farmland, the birds were extirpated entirely. In other areas, where forests were allowed to regenerate and pockets of big trees were left standing, ivorybills persisted, but in small and isolated populations. Begin Orgy of Consumption Act II. Many natural history enthusiasts, both rich amateurs and professionals, decided that the greatest thing one could possess was the skin of a dead ivorybill. The museum men used the excuse of needing the dead ivorybills as a record of their existence, but in reality collecting the skins of ivorybills was little different from collect-

A stand of girdled cypress taken in Florida around the turn of the twentieth century. Green cypress will not float, so trees were commonly girdled six to twelve months before they were cut to allow them to dry and become buoyant. (Courtesy of State Archives of Florida.)

ing baseball cards. The rarest card is always the one most earnestly sought, and a nearly extinct bird was irresistible to the collectors.

And what better time to obtain an ivorybill skin than when the birds were concentrated in a few remaining forest refuges? From about 1880 until about 1920 there was a slaughter of ivorybills that took out *every known individual.* I'm not exaggerating. In 1924 the great Cornell ornithologist Arthur Allen was led to a nesting pair of ivorybills in Florida by a local woodsman. It was the first verifiable ivorybill record to reach Allen in years, and as far as he knew, these were the last two ivorybills on earth. And yet, it was only with some regret that Allen "refrained" from collecting this pair himself. Instead, he spent a week studying their behavior. But when he and his wife left the area, collectors came to the nest and shot the birds.

With the shooting of the pair of ivorybills in Florida in 1924, most ornithologists and wildlife biologists thought that the Ivory-billed Woodpecker was extinct in the United States. In 1932, Mason Spencer, a country lawyer and member of the state legislature in Louisiana, mentioned to Armand Daspit, director of Louisiana's Fur and Wildlife Division, that ivorybills still lived in the woods along the Tensas River near his home in Madison Parrish. Legend has it that Daspit and his deputies at the state wildlife office in New Orleans retorted that he and the other hicks in Madison Parrish wouldn't know an Ivory-billed Woodpecker if it flew down and bit them in the ass. To drive his point home, Daspit wrote up an official collecting permit that would allow Spencer to shoot as many ivorybills in Louisiana as he wanted. I'm sure Daspit thought it was quite a joke—like giving someone official permission to trap leprechauns. I would love to have seen the look on Daspit's face when Spencer walked into his office a few weeks later and threw a freshly shot ivorybill down on his desk.

By shooting an ivorybill, Mason Spencer proved to a skeptical world that the Ivory-billed Woodpecker hung on in the northeast corner of Louisiana. Daspit immediately banned the shooting of ivorybills in Louisiana, revoking Spencer's permit. It didn't seem like such a joke any more. He also ordered a state game warden in the area, J. J. Kuhn, to make sure that no more ivorybills were shot. The woodpeckers were safe. Their forest, sadly, was not. The virgin forest in which these ivorybills had been rediscovered, an 81,000-acre stand of mostly virgin timber along the Tensas River, was owned by the Singer Sewing Machine Company, which is why the forest is usually referred to as "the Singer Tract." The Singer Company leased logging rights to the forest to the Chicago Mill and Lumber Company.

When Arthur Allen got news of the Ivory-billed Woodpecker population in Louisiana, he began making plans for an expedition into the area to study its breeding biology before the species was gone again. Allen's decision to go to Louisiana to study ivorybills would establish Cornell University as the world center for ivorybill research, and in a very real sense, it laid the groundwork for the rediscovery of ivorybills in Arkansas by Cornell scientists sixty years later.

In 1935, Allen visited the Singer Tract with George Sutton, his faculty colleague, James Tanner, a new graduate student, and Paul Kellog, a sound technician. The expedition was a stunning success. In April 1935 they found

Arthur Allen watching a breeding pair of ivorybills in the Singer Tract in 1935. (©Cornell Laboratory of Ornithology.)

a nest and captured the only motion pictures of Ivory-billed Woodpeckers ever made until, perhaps, David Luneau videotaped a fleeing ivorybill along the Cache River in Arkansas in 2004 (I'll return to David Luneau's video later in the book, and from this point on, I'll refer to this tape as the "Luneau video"). Perhaps most importantly for future ivorybill searchers, Allen's group made clear sound recordings of the kent calls—the diagnostic call note—of ivorybills.

James Tanner returned to the Singer Tract and spent the better part of two years, from 1937 to 1939, recording as much detail as he could about the life history of the Ivory-billed Woodpecker. Tanner's study, which he submitted as his Ph.D. dissertation at Cornell, was published as a book, *The Ivory-billed Woodpecker*, in 1942. It is undoubtedly the most widely read doctoral dissertation in the history of ornithology.

As Tanner was completing his study of Ivory-billed Woodpeckers, the Singer Tract was shrinking. The Chicago Mill and Lumber Company began exercising its logging rights in the late 1930s, and bit by bit the last intact swamp forest in the Mississippi River valley was cut. Some effort, including an appeal to Franklin Roosevelt by the Cornell Laboratory of Ornithology and the National Audubon Society, was made to stop the cutting of the

James Tanner helping with sound recording as part of the 1935 Cornell University expedition to the Singer Tract. (Photograph by Arthur A. Allen, ©Cornell Laboratory of Ornithology.)

Singer Tract and to save the ivorybill, but this effort proved futile. The entire Singer Tract was cut by 1944, which, not coincidentally, was the last year in which an ivorybill was reliably reported in the area. With the Singer Tract, it seemed, went the ivorybill.

Tanner not only studied ivorybills in the Singer Tract, but he also surveyed the entire range of the species for remaining birds. In his dissertation, Tanner all but declared the species extinct. He identified a few isolated forests such as the Apalachicola River basin in Florida and the Santee Swamp in South Carolina where a very few Ivory-billed Woodpeckers remained in 1939, but he wrote that all of these areas were being or would soon be logged. Tanner's assessment indicated that the ivorybills were doomed. The Ivory-billed Woodpecker was never listed as extinct by the U.S. Fish and Wildlife Service only because at a meeting of the Ivory-billed Woodpecker Advisory Committee in 1986 Jerry Jackson cast a cautionary vote in opposition to Tanner and Lester Short, who were ready to pronounce the bird extinct. Despite the fact the Ivory-billed Woodpecker remained on the list of endangered species, the accepted status of the species by the ornithological community through the last half of the twentieth century was "almost certainly extinct." In his widely acclaimed *Sibley Field*

Guide to the Birds, published in 2000, a comprehensive treatment of extant birds of North America, David Sibley did not mention the Ivory-billed Woodpecker.

Despite the proclamations of many of my fellow ornithologists, and for no particular reason other than that I thought that the world was still a pretty vast place, I always believed that Ivory-billed Woodpeckers might persist in remote swamp forests. Each spring in my ornithology lectures I would ask the students: "What's the difference between Bachman's Warblers and Ivory-billed Woodpeckers that makes Bachman's sure to be extinct but gives hope that ivorybills will be found?"

Blank stares.

"Bachman's Warblers were Neotropical migrants that moved down the Florida peninsula to their wintering area in the Caribbean. Even if Bachman's Warblers bred in the most remote swamps in North America, they had to reveal themselves each spring and fall. If they still existed they were certain to be seen in migration by birdwatchers. They had not been seen in migration in decades and were, beyond reasonable doubt, extinct. Ivory-billed Woodpeckers, on the other hand, need never leave their remote swamps and might live undetected indefinitely."

In my time in Alabama I had come to realize there were immense regions in the South (thousands of square miles) that are never viewed by anyone with book knowledge of birds, and where local people who might see something interesting would almost never have the inclination to report what they saw or be listened to if they did. The call from the man in Geneva County made me start to dream of conducting my own search for Ivory-billed Woodpeckers in the swamps of southern Alabama. But I had a research program to establish, small children to raise, and a million reasons not to go searching for ivorybills. "Some day," I told myself, "I'll mount an expedition."

In April 2005, ten years after the phone call from the man in Geneva County, having spent exactly zero time in pursuit of Ivory-billed Woodpeckers and really only thinking about them each spring when lecturing my ornithology class, I was stunned by the announcement that Ivory-billed Woodpeckers had been found in Arkansas. This was not just another tale from the swamp. This was an announcement in the most prestigious scientific journal in the country, *Science,* by perhaps the world's most distinguished institute for bird research, the Cornell Laboratory of Ornithology. The lead author on the rediscovery paper was John Fitzpatrick, who was past president of the American Ornithologists' Union, the top North American society for professional ornithologists and a recipient of the Brewster Medal, the Nobel Prize of ornithology. At a press conference at the Department of the Interior in Washington, D.C., Fitzpatrick announced that at least one Ivory-billed Woodpecker had been found along the Cache River in Arkansas. He explained that his search team had logged sightings and other evidence, but that the proof of the existence of ivorybills was a video of the bird. The Laboratory of Ornithology press release stated that the video was the first conclusive proof of an ivorybill in more than sixty years.

The news of the discovery of an Ivory-billed Woodpecker in Arkansas caught me, as it caught many birders and nature enthusiasts, completely off guard. It was extremely exciting, and I had not had the slightest hint of the discovery until the day of the press conference in April 2005 officially announcing the sighting. I was as far out of the Ivory-billed Woodpecker loop as a professional ornithologist could get. I knew several of the ornithologists involved in the Arkansas search, but I was not one of those "in the know." In the twelve hours following the announcement, I must have had half a dozen e-mails from birder friends and biology colleagues, half jokingly-half seriously asking: "Want to go to Arkansas?" I would write back jokingly: "Sure let's go," but I was thinking: "Not even if you handed me a plane ticket." Arkansas and the birder circus that I imagined had been created there was the last place on earth I wanted to be. Nonetheless, the announcement of the discovery of an Ivory-billed Woodpecker in Arkansas rekindled my desire to search southern Alabama for signs of the species. It was time to take my long-delayed trip into the swamp.

A Most Improbable Discovery   2

The spring of 2005 was a busy period in my ornithology lab at Auburn University. It was all I could do to supervise the seven graduate students, three postdoctoral fellows, and three field technicians working full time on half a dozen different research projects. This huge lab group included Tyler Hicks, working for me as a temporary technician doing bird point counts. Tyler is among the best birders I've ever met. As a matter of fact, in many ways, in the spring of 2005 he was already a better and more experienced birder than I was. This might not seem like an outlandish statement until you consider that in the spring of 2005 I was a forty-four-year-old professional ornithologist, and Tyler was twenty-one.

Tyler has that rare combination of keen eyes, extremely sensitive ears, a photographic memory, and an intellect that allows him to assimilate a vast store of information about birds. In the field he is simply phenomenal. I consider myself an excellent birder. Bird identification is my primary pastime, and one of my best professional abilities. When I take less advanced birders into the field, they rave about my skills. When I go into the field with Tyler, however, I'm always a step behind. He invariably spots birds faster and calls out the ID of birds, always correctly, while I'm still focusing my binoculars. The way I feel when I'm around him must be the way that average pro basketball players felt when they were on the court with Michael Jordan. I can imagine them muttering to themselves: "I really am a good basketball player—but that guy is superhuman."

*Tyler Hicks showing
his skills at wrangling
armadillos. (Photograph
by Brian W. Rolek.)*

Tyler's interest in natural history and birds brought him unwanted attention at his public high school in rural Kansas. Before his junior year, he got fed up with the bullying and teasing by his classmates, and he fled home and school. At sixteen he was already an avid birder when he decided his time would be better spent seeing the birds in new regions of North America rather than fighting to keep his lunch money. So he set off on his own with less than $50 in his pocket, and over the next several years he meandered his way around North America, hiking or hitchhiking from south Texas to Alaska to Quebec. Although he was only a teenager, he supported himself at a subsistence level as a technician on various research projects, banding or counting birds. By the time his name surfaced in a pile of applications for a bird technician job that I advertised in 2003, he was a twenty-year-old veteran field technician. Despite being so young, he was by far the most experienced field ornithologist who applied for the job, and I hired him.

Tyler and I hit it off from the start—our mutual love of birds, birding, and bird-related adventures dissolved our differences in age and backgrounds. Tyler has a maturity that far exceeds his age. Because he had become self-sufficient at such a young age and gained so much experience as a field ornithologist, when I talk with him it is hard to remember I'm not interacting with a forty-year-old colleague. Tyler hung out with the grad student group at Auburn, even dating a grad student, despite the fact that grad students were seven or eight and even ten or fifteen years older. From the day he arrived at Auburn, Tyler became my right-hand man for all of my bird/habitat studies, essentially running this large research program when I was busy with other projects.

For two years, at the end of each field season, Tyler would say his goodbyes two proclaim that it was absolutely the last year he would work at Auburn. There were too many fascinating places in the world to visit for him to keep coming back to boring Auburn doing point counts in the tiny and uninteresting Tuskegee National Forest. "No way am I ever doing counts in that trashy little forest again," he would proclaim. Despite his protests, though, each spring he was back. He had found a second home in Auburn, and he knew I needed his help to keep my projects going.

In the summer of 2005, Tyler was doing endless point counts in Tuskegee National Forest, the smallest, possibly most highly disturbed national forest in the country, as part of a study of the ecology of encephalitis viruses. Tuskegee National Forest had been created out of abandoned cotton fields at the end of the Great Depression in an effort to stabilize soils, and although it had some nice bottomland forest, mostly it was a big stand of loblolly pine. For Tyler it was boring work, and he was looking for some outlets for his birding energy. He, too, was captivated by the news of Ivory-billed Woodpeckers in Arkansas. I told him about the report of an ivory-bill in Geneva County I had gotten a decade earlier, and we started talking about a kayak trip down the Pea River to see if we could at least find Ivory-billed Woodpecker habitat. We asked around to see if anyone wanted to go with us, and another summer technician, Brian Rolek, immediately said yes.

Brian had only recently completed his undergraduate program at the University of Massachusetts. He was an engineering refugee, having bailed out of an electrical engineering career track to pursue his growing passion for natural history and ornithology. He was still quite green as a birder, barely beyond a beginner in his experiences and skills at bird identification.

*Brian Rolek investigating a large cavity in a
fallen tree. (Photograph by Geoffrey E. Hill.)*

But what Brian lacked in experience, he more than made up for in enthusiasm. The summer of 2005 was his second year assisting my grad student
Mark Liu with his doctoral research on Eastern Bluebirds. Brian was becoming a legend as a field assistant—not just willing but seemingly enthusiastic about twelve-hour days in the field, often for several weeks without
a break. He was absolutely dependable, careful, and accurate with all the
many tasks he performed—no wonder Mark wanted him back so badly
after his first field season.

So Tyler, Brian, and I made plans for a weekend kayaking trip down the
Pea River. We were all busy with our respective projects, and this was just
to be an overnight trip to see the river and get out in the nice spring
weather. None of us imagined that we had a snowball's chance in hell of actually finding an Ivory-billed Woodpecker. The Geneva County sighting
had been ten years earlier (talk about your cold trail) and I had gotten no
specific location, just along the Pea River somewhere in Geneva County.
We had another excuse for the kayak trip, though. Rick West and the Alabama Ornithological Society were working on the state's first breeding-
bird atlas, and we intended to keep track of the birds we encountered along
the Pea River for the atlas project.

We took my two kayaks and borrowed a third and left Auburn around 6 A.M. on Friday, May 20. We dropped off a vehicle at our take-out spot, and got to our put-in spot around 10 A.M. It was getting warm as we shoved off from our launch site. Barn Swallows swooped back and forth to their mud nests under the County Road 17 bridge as we paddled through the clear, flowing water of the Pea River. We moved downstream, assessing the forests on each bank of the river, and it was soon obvious that the heat and our late start were not the only obstacles to finding an Ivory-billed Woodpecker that day; lack of big trees and absence of swamp forest would do the job just fine. For the entire length of the Pea River that we floated, the banks were either steep and eroded or sandy. The river corridor was nearly completely forested, but we found no swamp forest. There were widely scattered huge trees, but the forest was generally second growth and not very old. In many places the native hardwoods had been removed and replaced with neat rows of slash or loblolly pine. It was a pleasant five-hour float with Swallow-tailed Kites overhead at one point, and Swainson's Warblers and American Redstarts serenading us down the river, but we saw no habitat that gave us the least hope for ivorybills.

We still had the entire next day to continue the search, and the initial plan had been to search another section of the Pea. But as we pulled our boats out of the river and talked about what to do next, Tyler said that on the aerial photos the other sections of the Pea looked even less promising, with less forest. The naysayers, it seemed, were right. The southern swamp woods were thoroughly and completely trashed. It didn't matter how remote and little-birded an area was if it didn't have old-growth swamp forest, which everyone agreed was essential habitat for ivorybills. As I was rambling on about the lack of promising habitat, Tyler made an offhand comment about how the forests got so much better across the state line in Florida along the Choctawhatchee River (pronounced Choct-a-hatch-ee), which the Pea joined just 2 miles downstream from our pullout.

"It's too bad that we can't search there," Tyler commented.

"Why can't we search the Choct-a-wat-chee?" I asked, butchering the name. (Throughout the first few months of our project we struggled to remember how to pronounce the Indian name for this river. More than once we wished we had found woodpeckers on the Pea just to make the pronunciation of the discovery site more manageable.)

"I don't know. You were always focused on Alabama swamps. I thought you wanted to stay in Alabama and do atlas work," Tyler replied, remind-

Brian Rolek paddling past an eroded bank along the Pea River on the Alabama/Florida border. (Photograph by Geoffrey E. Hill.)

ing me that I had always talked of rediscovering Ivory-billed Woodpeckers in Alabama and was following up on an Alabama sighting.

"Well, we've searched Alabama. There's no ivorybill habitat here," I ex-aggerated. I hadn't given up on finding ivorybills in Alabama, but this part of the Pea did not look promising. "Let's go to Florida."

"Really?" Tyler responded, like a kid whose dad had just said he could go to Disneyland. Tyler loves birding and exploring, and the prospect of better swamp forest was all he needed to recharge his optimism. "But we don't have any maps."

"We'll buy some on the drive south," I suggested.

One of the reasons that Tyler had looked at photos of the Choctaw-hatchee River when he was planning our Pea River float is he had run into an editorial by John Fleming in the May 4, 2005 edition of *The Anniston*

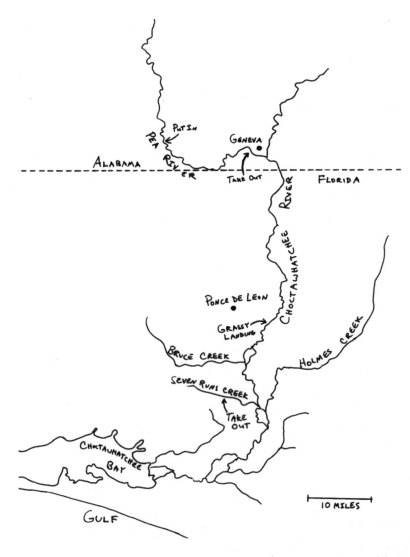

PUT IN
PEA RIVER
GENEVA
ALABAMA
TAKE OUT
FLORIDA
RIVER
CHOCTAWHATCHEE
PONCE DE LEON
GRASSY LANDING
BRUCE CREEK
HOLMES CREEK
SEVEN RUNS CREEK
TAKE OUT
CHOCTAWHATCHEE BAY
10 MILES
GULF

A map of the Pea and Choctawhatchee rivers with our put-in and take-out spots indicated. See the map on page 81 for details of the Seven Runs Creek area.

Star, the paper for the small town of Anniston, Alabama. Fleming had recounted tales that his grandfather told about ivorybills along the Choctawhatchee River. Of course, this same grandfather also told tales of a monkey man and a hog monster that disemboweled wild pigs, so it hardly seemed like the most credible sighting. Nevertheless, a big swamp forest and legends of ivorybills were more than we had in Alabama.

So around three in the afternoon we drove south on Highway 87, watching it change to Highway 81 as we crossed the state line into Florida. Tyler remembered from his brief scan of aerial photos in the Auburn library that the forests got broader and more mature south of I-10. My road map of Alabama included just enough of Florida to show that Ponce de Leon was the town immediately south of I-10 along the Choctawhatchee River, so we drove toward Ponce de Leon.

No trip could have been conceived more whimsically. All we had were Tyler's recollections from glimpses of fifteen-year-old aerial photos that the swamp forest along the Choctawhatchee River got broad and looked pretty mature south of I-10. Because I had talked only about searching Alabama, Tyler had prepared for an Alabama float with maps and ideas for put-ins and take-outs and even places for us to stay for the night. Now we had shifted 40 miles south, and we were guessing. We pulled into the tiny town of

Brian (left) and Tyler (right) at the start of our float down the Choctawhatchee River on May 21, 2005.

Ponce de Leon about 4 P.M. and got some snacks and, most importantly, a map. Unfortunately, all they had at the gas station was a map of Walton County that looked a little rough (an impression that we verified the hard way the next day). A boat launch labeled "Grassy Landing" was marked on the map just three miles or so south of the edge of the town, so we decided first to make for that.

Grassy Landing was more obscure and harder to find than the map made it appear. We ended up driving down an unmarked two-track road through pine plantation expecting any minute to run into an irate land-owner, when suddenly we were at the river. There was a slightly wider point in the track where a car could be turned around and the road ran down into the water—apparently this was the landing. It was fine for us and our kayaks. We would come back at dawn and start our trip. Pine plantation with rows of slash pine saplings came right down to the river here, but there were a few young cypress trees with Spanish moss at the boat launch (virtually the first cypress we'd seen on our trip), so at least we felt as if we were on a southern river. It hardly looked like ivorybill habitat, though.

We next headed south on Highway 81, the main north-south artery that parallels the river on the west side, looking for a take-out spot. About 8 miles south of Grassy Landing there was a newly graveled road with a sign, TILLEY LANDING RECREATION AREA, that led east toward the river. This landing seemed to be just the distance downstream that we wanted for our trip the next day. The road looked passable, so we went down. Almost immediately we were in the type of old-growth swamp forest that the literature suggests should be the habitat of Ivory-billed Woodpeckers. A new road ran on an elevated bed through a forest that was clearly flooded each spring. Large oaks and spruce pine grew from the drier areas, and from the lower sloughs, many of which had standing water, impressive tupelo and cypress grew. The forest looked quite mature—not virgin, but not cut for decades and perhaps not for a century. And the newly graveled road went on and on for about a mile through the forest before finally ending at a small canoe launch on a small finger of Lost Lake ringed by huge cypress. This slough looked exactly like the sort of swamp forest where we thought ivorybills would live. The problem was that this landing was not on our map, and we were clearly not at the river.

That's when we lost our minds and decided we would be able to find this slough and the little landing from the river the next day. As I look back on it now, I have to wonder what we could possibly have been thinking. I guess we were excited about finding swamp forest with Ivory-billed Wood-pecker potential and had yet to fully comprehend just how large this for-ested river basin was. "This must connect to the river a little way to the east," I concluded, foolishly assuming that any landing would be constructed to provide access to the river. Tyler took out his Global Positioning System (GPS) unit, marked the spot, and agreed that the river was about 2,000 feet away to the east. GPS units give your location on the earth's surface to within a few meters and can track your route from any starting point. They are essential equipment when exploring broad tractless swamps.

"Good, we'll pull out here at the end of our float tomorrow," I declared. In fading light we drove back to Ponce de Leon and spent the night crowded into a room in the only motel in town.

We got an early start the next morning, Saturday, May 21, 2005, drop-ping my Honda CR-V at Tilley Landing and putting our boats in at Grassy Landing a few minutes before dawn. It was a gorgeous day with no wind and clear blue skies. Mist rose off the water as we pushed off through a patch of willow sprigs into the main channel of the Choctawhatchee River and drifted south. The current was stronger than I had anticipated. I was expecting a slow-flowing blackwater river, and instead there was a steady 3 mph current. It was also immediately obvious that this river had more ex-tensively forested borders than the Pea River in Alabama, but the first half-mile of our float was not particularly impressive. Soon after we launched, a steep, eroded bank loomed to the west with rows of pine planted to its edge. I was mostly focused on enjoying the kayaking; any real thoughts of ivory-bills had vanished the day before. About a mile below our launch site, we took a fairly large and fast-flowing side channel to our right. We would later learn that this channel is called the "Roaring Cutoff" because of its swift current. A short way down the channel we had to crash through the branches of a tree that had fallen completely across the channel. There is no way a motorboat or even a canoe could get into this channel unless the boatman was carrying a chain saw. For our kayaks, though, it was a minor obstacle.

As we rounded a bend, the current let up a bit, and I just floated along, noticing that the forest was now quite impressive with very large oaks and a few huge cypresses. Also, the banks were shallow and in many places the river overflowed into the forest. This was the sort of mature, frequently flooded bottomland forest that was reported to be ivorybill habitat. As I drifted along, craning my neck to look into the canopy high above my head, Prothonotary Warblers sang from both sides of the channel, and a Red-shouldered Hawk screamed at us as it circled overhead.

Tyler and Brian were 100 feet behind me discussing how nice the forest had gotten as I approached the mouth of a creek. I was turning my kayak into the creek channel when I heard very loud knocking coming from the big trees nearby. I looked back at Tyler and Brian and half shouted, half whispered, "Loud knocking!" I pointed at the creek channel and paddled in. I don't think Tyler or Brian heard me, but they followed, and we were soon stopped side-by-side staring at the canopy of the forest in the direction of the heavy knocking. It seemed to be slower and much heavier than any sound I've heard a woodpecker make before. To me, it sounded like a person banging a heavy mallet on a solid wooden beam. It seemed to be a foraging rap because the knocks were irregular. The bangs were considerably slower and much heavier than the foraging taps of Downy, Hairy, Red-bellied, or even Pileated Woodpeckers, which I hear regularly. As we stared at the dense canopy, the knocking seemed to move farther into the forest, and Tyler, Brian, and I followed, moving a hundred feet or so down a small channel in the flooded forest until the sound seemed to come from high in the big trees right in front of us. The knocking had gone on with a few interruptions for maybe 10 minutes by this point, and we ended up with Brian's kayak and my kayak side by side and Tyler twenty feet or so ahead and to our left. I could have easily put my hand on Brian's shoulder. We were right next to each other. I make this point because just then as I was scanning left, Brian said, "Did you see that?"

"See what?" Tyler and I answered in unison, clearly having seen nothing.

"The bird that flew off, right from where the knocking was. It was very white," Brian answered, a bit excited.

"What do you mean very white?" Tyler asked, not yet willing to accept that he might have just missed something special.

"It had a lot of white in its wings," Brian replied, still trying to figure out what had just flashed in front of him.

"Just in the wings?" Tyler asked, almost dreading the answer. We had flushed several White Ibises as we pulled into the creek.

"Yeah," Brian answered, "it was big and mostly black with a lot of white flashing on the top and bottom of the back of its wings."

"Damn," Tyler and I responded in unison.

I had missed that bird by about 10 degrees of my field of view. If I had been looking forward, or certainly to the right, I would have seen it. But, of course, I was looking to the left and missed it completely. It was about 7:30 by this time and things were certainly getting interesting. We had found amazing Ivory-billed Woodpecker habitat, heard knocking unlike anything I'd heard before, and Brian had seen a large bird with a white trailing edge to its wings fly away from the spot.

We proceeded up the creek, which Tyler informed us was Bruce Creek on his GPS, and we became more and more impressed by the forest. It was not a virgin forest—we could see the stumps of the big cypress that had been cut out. But it was a big, mature woods that obviously flooded regularly. The forest floor was covered in mud, and there were recent water marks 3 feet up the trunks of all the trees. Oaks 3 and even 4 feet in diameter towered 100 feet over our heads. And here and there were huge cypresses—perhaps too hollow or imperfect to have been cut. Most intriguing, on some recently dead and even some of the living hardwoods, there were areas of chiseled bark the likes of which Tyler and I had never seen before.

In his dissertation on the Ivory-billed Woodpecker in Louisiana, James Tanner wrote that 70% of the Ivory-billed Woodpecker foraging bouts he observed were of birds "scaling bark from dead trunks and branches to secure the borers that live between the bark and sapwood." He also used the term "peeling bark." "Peeled bark" was not a good description of the feeding signs that we were seeing. "Scaled bark" would have been a better description, but I think that the best term for what we saw is "chiseled bark." On the side of a tree there would be an area about 10 inches square where a section of thick, strong bark had been pried off the tree. Invariably, at the center of each small section of removed bark (about 2 inches square) within the larger feeding area, there would be a bore hole where a large beetle larva had been living. On the freshest of the feeding sites, fibers of wood and bark would bound the scar, and I was unable to budge the bark at the edge of the chiseled sections with my fingers. If I had wanted to remove a similar section of bark with a sharp chisel and sledge hammer, I could have done so only with considerable effort. I had seen woodpecker-feeding trees my entire life, but I had never seen anything like these chiseled sections from the strong bark of trees. These bark scalings had clearly been made by powerful birds. We photographed several of the most impressive chisel marks.

After about 30 minutes of exploring Bruce Creek with no more knocks or other suspicious sightings or sounds, we floated back out to the river. We hugged the west bank of the Choctawhatchee River as we floated south, and within a few hundred meters a bay opened to our right. We paddled

into the small bay and immediately saw a large dead tree with three huge oblong holes stacked one above the other. They seemed larger than the cavities that Pileated Woodpeckers cut, of which I've seen many,

Tyler and Brian paddling in Bruce Creek just minutes after Brian saw an ivorybill fly from the canopy. (Photograph by Geoffrey E. Hill.)

but we were trying to remain objective. We moved to the north end of the bay and then up a narrow channel into a magnificent stand of flooded cypress. These trees were big, some more than 6 feet in diameter with knees projecting 6 feet above the water, and the stand went on for a long way. I was amazed that a stand of cypress like this still existed. I had gotten the impression from the ivorybill literature that every large cypress in the South had been cut by the early twentieth century. Yet here we were in a magnificent gallery of cypress, some trees having been removed, but still containing numerous ancient giants.

We weaved our way through the flooded cypress for about 20 minutes. When we had gone about as far into the stand as the water level would allow and turned around, I heard a clear double knock off to my right (west). At this time, I had never heard a recording of a double knock, and although I had seen *Campephilus* woodpeckers in Costa Rica and Brazil, I had never heard them knock. What I heard from this cypress stand was the same sort of heavy knocking that I had heard an hour before on Bruce Creek, but this time it was just two knocks. The second knock immediately followed the first and was slightly softer. I waited for a minute, heard nothing, and then moved up to Tyler and Brian, who were about 100 feet in front of me. "Did you guys hear that double knock?" I asked. They hadn't and, as it turns out, none of us heard another double knock for the rest of the summer or fall.

For the remainder of the day we explored what seemed like innumerable creeks and little channels that took us into the forest on both sides of the river. Everywhere we went there was an impressive flood forest with huge oaks and cypresses. I was staggered that this place had never gotten even a mention in any discussion of potential places to look for Ivory-billed Woodpeckers. I had seen several lists of the places most likely to harbor ivorybills posted by ornithologists, birders, or bloggers, and the Choctawhatchee River basin was on none of those lists. It was not mentioned by Tanner among the places that he checked for ivorybills in his continent-wide survey in the 1930s, nor by Jerry Jackson during his ivorybill searches in the 1980s and 1990s.

When various authorities had stated that there were no large, mature swamp forests left in the Southeast, I had believed them. But they were wrong. The forests of the Choctawhatchee River are mature, very extensive, and virtually inaccessible. This vast forest is not virgin, but it was never clearcut, and most sections have been largely undisturbed for seventy or eighty years. The seasonally flooded portion of this forest has an average width of about 1.5 miles and runs for at least 40 miles from I-10 to Choctawhatchee Bay. That's approximately 60 square miles (almost 40,000 acres) of inaccessible swamp forest just along this lower section of the main channel of the Choctawhatchee River. There is also extensive swamp forest along tributaries like Holmes Creek and in the numerous bays scattered across the adjacent countryside.

Everywhere we went we saw what we thought were signs of Ivory-billed Woodpeckers—chiseled bark and large cavities—but we didn't hear or see anything else that suggested an actual Ivory-billed Woodpecker the rest of the day.

By about 4 P.M. we were nearing our take-out point. Tyler was watching the GPS and told us we had just passed within a quarter mile of my car to our west, but we saw no channel of any sort headed that way. Ahead the river went around a large bend. We figured the channel leading to our vehicle must be around the bend, so we floated away from the car. We ended up floating nearly two miles away from the car with no sign of a channel headed that way. Finally, we neared the entrance to a large bay at the mouth of Seven Runs Creek. We had gone too far. We started the laborious task of paddling upriver against the considerable current. At this point we had been in our kayaks since 6 A.M., about 10 hours of paddling without a break (we had eaten our meager lunches while we paddled), and I was getting tired. We paddled upstream against a stiff current for about 30 minutes, but it wasn't clear where we were headed. Finally we paused to discuss our options. Even though we were well south of the car now, we decided to push into the flooded forest and try to make our way to Lost Lake. We found a small channel flowing through the woods headed more southwest than west (and we really wanted to go northwest from this spot), but we were tired of paddling against the current, so we took it. Ten minutes later we popped out in the bay near Dead River Landing at the mouth of Seven Runs Creek as a fishing boat zoomed past. We had little choice—we paddled into the bay and headed toward the northwest corner, now far from the car.

Despite having quite a bit of boat traffic (by far the most we had seen all day), this area had many huge cypresses and looked like great ivorybill habitat. At this point, though, we were hunting for our car more than for woodpeckers. It was now after 5 P.M. The sun would set a little after 7, and it would be completely dark before 8 P.M. We couldn't keep circling around forever. We found no channel running in the right direction from the Dead River area, so we decided our landing must somehow be up Seven Runs Creek.

The good news was that with heavy spring rains Seven Runs Creek had enough water for kayaks. The bad news was that it also had a substantial current. We paddled past Dead River Landing waving to the people. This is a major access point for the river, and on this Saturday there were six or seven cars, several tents, and a few campfires going. We wished we had parked there. Our car was now about 2 miles north of us, and we seemed

Tyler paddling in a stand of large cypresses near Dead River Landing. (Photograph by Mark Liu.)

to have no way to get there. Still we thought there might be a channel to Tilley Landing ahead, so we paddled due west up Seven Runs Creek.

A few hundred feet above the landing, we passed a man and his hound dog in a john boat moored next to a huge cypress fishing for crappie.

"Is Tilley Landing this way?" Tyler asked him.

The man looked puzzled. "Don't know about no Tilley Landin'. The highway's up yonder a ways." He pointed up the creek.

It was now about 6 P.M. and we couldn't fool around any longer. We decided the only sure way to our car was to paddle up Seven Runs Creek to the highway and then to walk the three miles to the car. I estimated we'd get back to the motel by around 10 P.M., if nothing went wrong. Tyler checked the GPS and told us it was 1.8 miles to the highway.

In my pampered suburban life, I've faced few serious physical challenges. In my younger days as a spelunker, I once had to literally drag my best friend out of a cave when he dislocated his shoulder in a remote passage, miles from the entrance. That had been a grueling and totally exhausting 10-hour ordeal. But our paddle up Seven Runs Creek at the end of that long day stands as my toughest physical struggle. We started this upstream paddle after about 12 hours in our kayaks, having already spent a good portion of the late afternoon paddling against the strong current of the Choctawhatchee River. I felt like my arms were spent when we started upstream, and it proved to be quite a hellish paddle. The channel was so narrow and shallow that it was impossible to get good deep strokes, so we were left with little choppy strokes against the relentless current. Worse, about every 200 feet a tree was down over the creek. For each fallen tree we had to get out, pull the kayaks up the steep muddy banks and around the tree, and start again. We did this over and over. It took us about 90 minutes of steady hard paddling and pulling to reach the highway. I have never been so glad to hear cars and see a strip of asphalt.

There is a little park, just a patch of grass and a small pavilion, where Highway 81 crosses Seven Runs Creek, and we pulled our boats out of the water and collapsed for a moment on the grass in this park. I'd been in the kayak for more than 13 hours, and the last 3 had been mostly spent paddling steadily against a brisk current where even a moment's pause would send you back downstream. I could barely lift my arms to swat the evening mosquitoes that were descending upon us.

We couldn't sit around for long. The sun had set, it was getting dark, and we still had to walk 3 miles to the car. We decided to leave Brian with the boats (I have no recollection of how this was decided), and Tyler and I set off down the highway. It was about 2 miles to the gravel turnoff for Tilley Landing and then about 1 mile to the landing. We figured we would be there in an hour. After the paddle from hell, our stroll down the highway was a welcome relief—my arms were not required to do anything but hang at my side. We walked along contemplating what we had seen and heard that day as cars occasionally whizzed past on the highway.

Just as we could see the gravel road leading to Tilley Landing about a half mile ahead, a pickup truck pulled over in front of us. We walked up and looked in the passenger window, and a man somewhere in age between Tyler and me asked, "You boys need a lift?"

"Yeah, thanks a lot, but we're going to Tilley Landing," I answered, figuring the man had offered to take us up the highway, not a mile off of it.

"T' where?" he asked, as puzzled as the crappie fisherman had been. Apparently no one who lives in this county called the new landing "Tilley Landing."

"The landing at the end of that road up ahead," I added.

"Oh, down that new road," he said, now understanding exactly where we were headed. He paused for a minute and then said, "OK. Get in. I'll run yas down."

Tyler immediately blurted out "I'll ride in the back," and jumped into the bed of the pickup. He later told me he expected the guy to try to rob us and thought it was best if we weren't both in the cab of the truck. I climbed in the passenger seat, being careful not to get mud in the man's truck.

"What're you boys up ta?" he asked as we headed down the road.

"We're from Auburn doing bird surveys," I answered. "We misjudged our canoe take-out and ended up at the Seven Runs Creek with our car down here. You really bailed us out."

"Ain't no problem," he responded with a nod of understanding as we started bumping down the gravel road. "It's easy t'get turnt around in the swamp."

Finally the headlights of his truck reflected off my car in the parking lot ahead. The man stopped his truck and I hopped out.

"I can't thank you enough," I said as I searched for my keys in one of my zipped pockets. "It would have been a long walk in the dark."

"Ain't nothin'," he replied. And then as an afterthought he added, "Say, you got a dollar?"

"A dollar?" I repeated, wondering if I should have offered to pay him for the ride. Generally in the rural South, folks help each other out, and offering money would be insulting. I try to help out stranded motorists whenever I can, and I would never ask for any payment.

"Yeah, I hate t'ask but I was fixin' t' get sompen t'eat an I'm short a dollar," he added rather sheepishly. Apparently he understood that the code of helping neighbors made payment unseemly.

"Sure," I responded, trying not to sound surprised by the request. "I'd be happy to pay you more for going so far out of your way."

"Nope. A dollar'd be fine if you're willin'," he said, still sheepish.

I handed him a dollar. "It's my pleasure. Thanks so much for the ride," I said as I got the doors to my car open.

"You boys be careful now," he called out as he drove off.

We piled our wet, tired bodies into my CR-V and zoomed back to

Brian, who had the kayaks drained and dejunked. It didn't take us long to load up the boats and go pick up our other vehicle.

The choices of restaurants are slim in Ponce de Leon (either "Sally's" or a little, greasy place that the motel clerk thought had no name). Even though we were exhausted, we decided to drive to DeFuniak Springs, the next exit down the expressway, and celebrate with a better meal. We ended up at a Mexican place, toasting our discovery with Corona. We had found an Ivory-billed Woodpecker and heard it double knock within 24 hours of the start of our search. If you only count time in decent habitat, we had found an ivorybill in the first hour of the first day of our search in a place no authority had ever mentioned as a good place to look. Not bad for three guys who had just taken up ivorybill hunting that weekend.

From Possible to Virtually Certain   **3**

Tyler, Brian, and I decided to tell essentially no one that we thought we'd found ivorybills along the Choctawhatchee River. At this point I was far from convinced, and none of us wanted to be associated with a false claim of ivorybills. As I mentioned before, I had a huge group of ornithologists from technicans to postdocs in my lab that summer, but we decided not to make an announcement. It wasn't that I didn't trust members of my lab. I just thought that keeping a secret like Ivory-billed Woodpeckers in Florida would not be easy, especially when everyone in the ornithological community was abuzz with ivorybill chatter.

The only lab member whom we did not keep in the dark about our discovery was Mark Liu, my doctoral student from Taiwan. Mark was living in a house with Brian and Tyler, and the three were close friends. There was no way that Brian and Tyler could live with Mark, exuberant about the ivorybill discovery, and not tell him. I was happy to bring Mark in on the search anyway because Mark is an extraordinary bird photographer. What I mean by "extraordinary" is that Mark has an uncanny ability to get images of distant, moving birds under poor light conditions. I've fallen into the bad habit when birding with Mark of yelling, "Mark, photograph that bird!" when we see a rarity.

I first witnessed Mark's digi-scoping prowess on a trip to Dauphin Island, Alabama, in 2003 soon after he arrived from Taiwan. We found a female Harlequin Duck, a very rare bird for Alabama, at the ferry landing. As she drifted away from us, I shouted for Mark to photograph her because I knew he had a digital camera with him. I was just hoping for a speck in the picture that might yield minimal diagnostic field marks when enlarged. Instead, Mark got several amazing shots of the duck well positioned and in focus through his spotting scope, despite the facts that it was an overcast day, the duck was 100 yards distant and mostly facing away from us, and she was bobbing in pretty strong waves. I was impressed.

On a subsequent trip after a hurricane, we found a storm-petrel pattering well out on Lake Eufaula—the first living storm-petrel observed inland in Alabama and a tough ID for me (Wilson's or Band-rumped or Leach's?). Again, under challenging conditions with a distant, constantly moving bird as the target, Mark got great shots. The storm-petrel photos turned out to be especially important because although I identified it as a bandrump in the field, several experts on tube-nosed seabirds thought the photos virtually clinched Wilson's. I gladly deferred to the seabird experts. Anyone who could get nice shots of a distant duck bobbing in the surf or of a pattering storm-petrel had a much better chance than me or the rest of our crew of getting a photo of a flying ivorybill. So Mark was in from the start.

After our brush with a possible (Brian would say certain) ivorybill on Saturday, Brian and Tyler decided to spend Sunday back on Bruce Creek. I had to get back to Auburn that morning—I'm divorced and Sunday is one of my days with my kids, Trevor and Savannah. I had to pick them up in Auburn at 10 A.M. and I couldn't be late. So I dropped Brian and Tyler off at Grassy Landing before first light with their truck parked downstream at Dead River Landing and headed home. They retraced the fun part of our route of the previous day, skipping the part where we got lost. They didn't see anything that suggested an ivorybill, but they did hear a clear kent call near the mouth of Bruce Creek an hour or so after dawn. They also explored more backwater areas and saw more large cavities and more trees with chiseled bark.

Brian and Tyler's Sunday float was made a bit more difficult and cut a little short because my green kayak, which Brian was now using, started leaking. I had put a strip of duct tape over a weak spot on the stern of the boat for the previous day's trip and I'd had no leaking, but the tape had apparently come loose while I was fighting my way up Seven Runs Creek. I forgot to mention the leak to Brian and Tyler. Brian was taking on water almost from the start and had to bail constantly all day to keep from sinking. I'm sure he thought of a few descriptive terms for me as he bailed water in the swamp; fortunately, I wasn't there to hear them.

Bruce Creek near where it meets the Roaring Cutoff during a period of low water in November 2005. Water levels commonly rise and fall by 6 or 7 feet along the Choctawhatchee River. The log near the center of the photo, which here rests 6 feet above the creek, was completely covered by water during the start of our winter search in January 2006. (Photograph by Geoffrey E. Hill.)

After that weekend we were faced with the deepening heat of summer, the drying of the swamp, and an explosion in the mosquito populations, which had not been bad on our discovery weekend. We knew that if indeed we had discovered Ivory-billed Woodpeckers, we had discovered them at the very end of their breeding season. In his description of ivorybills in the Singer Tract in Louisiana in *The Ivory-billed Woodpecker*, James Tanner states that searching for ivorybills is very tough in the summer and fall and that the only time he could reliably find birds was winter and spring. But we were too excited to just sit around and do nothing for six months. We had to get back down to our wetland wilderness and follow up on our tantalizing glimpses of May 21. Ivorybills don't vaporize in the summer; they just get a bit more secretive.

As the cowboy relied on his horse, so the ivorybill hunter depends on his kayak. Most of our searching was done from a seat of kayak. (Photograph by Geoffrey E. Hill.)

On May 27, one week after our first visit to Bruce Creek, Brian was in the middle of the busiest week of the year for the bluebird project. Mark was Brian's boss and slave driver and understandably wouldn't even consider letting Brian take that weekend off. But Tyler and I were free, so we headed down for a couple of days in the swamp. We went in one vehicle with the intention of paddling back to our put-in spot at the end of each day. That meant that we couldn't go farther downstream from Grassy Landing than Bruce Creek, and even then we would have a tough upstream paddle of more than a mile to get back to the car. We would later learn that it was a much easier paddle from Bruce Creek Landing up and down Bruce Creek, but at this point we knew little about the area.

I should digress for a moment and describe the kayaks that we used for all of these trips and for our winter and spring search. These were not sports kayaks made for running rapids and performing stunts and rolls. Nor were they the big marine kayaks with rudders like the one I had used to paddle in a fjord in Seward, Alaska, the previous year. These were recreational kayaks—short and wide and very light. They are the simplest single-person boat imaginable—just room for an adult to squeeze into with very little

Tyler escaping a downpour by using his kayak as an emergency shelter. (Photograph by Brian W. Rolek.)

gear behind the seat or between the legs. Some models have one or even two compartments with waterproof rubber lids, but I prefer the simpler and lighter boats with no compartments. My kayak weighs less than 40 pounds, is 9 feet long, and has no storage compartments. It is made of a durable plastic that seems virtually indestructible (I'm constantly crashing into rocks and logs or throwing the boat off the top of my car). The only reason my boat developed a hole in the stern is that I had repeatedly dragged it over concrete and other hard surfaces and wore through the plastic. The hole was easy to patch.

I can put my kayak on the roof of my car by myself. Before I got involved with the ivorybill search, I used it mostly for solo fishing excursions for striped bass and longnose gar in the Chattahoochee or Tallapoosa rivers. My days spent fishing turned out to be useful because I was very experienced and comfortable in my boat when we first went looking for ivorybills. Tyler also was an experienced kayaker; he had conducted bird surveys of mangrove forests on Guantanamo Bay, Cuba, using sea kayaks. The morning we shoved off into the Pea River was Brian's introduction to kayaking.

My kayak only draws about 3 inches of water when I'm sitting in it, so

I can paddle it down even a tiny channel or through a forest flooded with shallow water. It's short enough to turn around in very tight spots. Because it's so light, I can also "jump" logs and other debris in it quite well. As long as I can get the tip of the kayak up on the object blocking my path, I can pull myself and my boat up and over it, walking forward on my hands if necessary. In my tiny kayak I can go places that no motorboat or canoe could possibly go, and if necessary I can easily get out and carry it or drag it for long distances through the forest. Recreational kayaks can even be propped upside down to make an emergency rain shelter. They are the ultimate swamp-search vehicles. The same feature that makes them great swamp boats—a wide, flat bottom—makes them difficult to paddle in open water. I wasn't thrilled about a mile-long paddle up the river at the end of our day.

Tyler and I had beautiful weather again for this second trip into the swamp. We got on the river at first light and split up when we entered Bruce Creek just after sunrise. I spent the morning drifting around, listening and looking. I saw or heard nothing that morning that suggested an ivorybill, but it was a glorious day to be in the swamp. Bruce Creek in early June is a birdy place. Downy, Hairy, Red-bellied, and Pileated Woodpeckers are all common, although no more common than in the woods around Auburn. Prothonotory Warblers seemed to be everywhere, giving their loud "Tweet! Tweet! Tweet! Tweet! Tweet!" calls from stumps and snags projecting above the water of the creek. The canopy was alive with Red-eyed Vireos, with their endless monotonous calls ("Here I Am, Up in the Tree, I Can See You, But You Can't See Me")coming from every direction. White Ibises and Little Blue Herons moved up and down Bruce Creek, distracting me when I caught glimpses of them moving through the trees.

About 11 A.M. Tyler came paddling toward me from farther up the creek. He had a huge smile on his face.

"I saw the bird," he proudly announced.

"No way! An ivorybill? You saw an ivorybill?" I exclaimed. (I'm told I greet any surprising or interesting news with "no way." "No way!" I exclaimed when Brian first told me about this mannerism.)

I was a bit dejected that I had missed the bird again, but I was mostly relieved. Brian's sightings had left a huge lingering doubt in my mind. Brian did not have the birding experience to call a flying ivorybill with absolute certainty. Tyler, on the other hand, is the best birder I've ever been associated with. There is no one whose sighting I would trust more.

"Yes," Tyler answered. "I spent some time around a feeding tree hoping a bird might come back. I didn't see anything there so I started down a channel to Bruce Creek. There was an old deserted bridge blocking my path. I noticed some small bird activity so I got out of my kayak and started looking at the warblers in the trees. Out of the corner of my eye, I thought I saw a bigger bird. I turned and there was an ivorybill flying through the woods."

"Wow. Did you see it land?" I asked excitedly.

"No. It flew across my field of view for a few seconds, and then it swooped up into the canopy and disappeared," Tyler answered, moving his right hand in an upward swoop. "As it flew across in front of me I could see a bright white trailing edge to its wings on both the upstroke and downstroke."

"Was it a naked-eye look?" I asked. It didn't sound like Tyler had a chance to raise his binoculars.

"No. I got it focused in my binoculars. As it swooped into the canopy, I had a great view of its upper side through my binoculars. I could clearly see white lines running from the neck down each side of its back. Besides the bright white secondaries and those white lines, the whole plumage was jet black. I got a great look at the head, and it was black. It was a female."

"What about the bill, did you see that?" I asked, knowing that the bill was not supposed to be that conspicuous on a flying ivorybill.

"Not well. Right after the bird disappeared, I thought about the bill. I think it was pale, but that's the weakest part of my sighting. Mostly I focused on the distinct plumage traits," Tyler concluded.

"You're sure it was an ivorybill?" I felt like I had to ask.

"Absolutely positive. I had it focused in my binoculars about 70 yards away. It was a great look." Tyler was beaming.

Nobody collapsed. There were no tears of joy. No hysteria. Tyler was clearly excited. I was relieved. The best birder on our team had gotten a clear look at the bird. Our ivorybill discovery had just moved from possible to virtually certain. The only reason that I don't say absolutely certain is that Tyler had seen the bird alone and it was a species he was very intent on finding. I couldn't entirely rule out that even a great birder like Tyler could have deceived himself into seeing the bird in the same way a deer hunter will turn a human into a deer as the excitement of a hunt overtakes him. I had birded with Tyler a lot, though, and he had never hallucinated a bird before. It was infinitely more likely that an Ivory-billed Woodpecker had flown in front of Tyler than that Tyler had seen field marks on a bird that didn't exist.

Tyler, Brian, and Mark made another trip to the site in late June. They spent one day exploring the Lost Lake and Dead River area a few miles to the south of our sighting locale on Bruce Creek. They reported that the forest was mature throughout this area, with some humongous cypresses and stands of huge oaks. They also spent a morning in the Bruce Creek area. As with all trips to the swamp forest, they saw lots of feeding trees and cavities, but on this trip, nobody in the group saw or heard anything that suggested an ivorybill.

In mid-July, Tyler took off for Kansas to get ready for his new school year. Brian finished up his work on the bluebird project in Auburn at the end of July, and he and Mark made one last trip to the Bruce Creek area. The fact that Brian was willing to spend two full days in this swamp in late July speaks to his enthusiasm for the project. Late July on the upper Gulf Coast has to be experienced to really be appreciated. To say that it is hot and humid in mid-summer is like saying that Fargo is cold in the winter. The north Gulf Coast in July is oppressively, suffocatingly hot, and it never cools down. It is common to wake to a predawn temperature of 80°F with 100% humidity and then watch the thermometer climb to the upper nineties with no drop in humidity. On most days there is no breeze. The heat envelops you like a hot, damp blanket. As bad as the heat can be, it is bearable if you can avoid exertion and wear as little clothing as possible. The beach in the shade is the place to be. But the thing that makes the swamps of the Gulf Coast truly unbearable in the middle of summer are the mosquitoes—endless swarms of blood-sucking, encephalitis-carrying mosquitoes. The only way to escape the mosquitoes is to completely cover your body, and it's too hot for clothes. If you need more reasons to stay out of the swamp in the summer (besides the snakes), Ivory-billed Woodpeckers become cryptic in the summer, giving few vocalizations or double knocks, and so they are nearly impossible to locate. For Brian to venture forth under these conditions, he had to be hooked.

Brian was in the Bruce Creek area for two full days during the last week of July, and he called me on his cell phone an hour or so after he got out of the swamp.

"Geoff, I saw it again," he said in an excited voice.

"No way. The ivorybill? You saw the ivorybill?" I exclaimed, amazed that he had been able to find the bird in the summer.

"I saw two," he added, clearly pleased with his effort.

"Two! You saw two birds?!"

An ivorybill in flight over the Singer Tract in 1935 showing a distinctive silhouette with long wings, a long neck and head, and a long, pointed tail. (Photograph by Arthur A. Allen, ©Cornell Laboratory of Ornithology.)

"Yeah. I saw nothing on Saturday [the first day of his trip], then on Sunday I didn't see anything in the early morning. It was really hot and the mosquitoes were terrible. If I stayed in the main creek channel and kept moving they weren't quite so bad, so that's what I did. Around 10:30 I was floating along, looking up, when an Ivory-billed Woodpecker flew right over my head."

"Did you see the white on the trailing edge of the wing?"

"No, it was getting sort of dark. Thunderstorms were moving in, and against the gray sky I couldn't see any colors on the bird. The shape was perfect, though. Just like that picture on the Cornell web page."

The Laboratory of Ornithology web page devoted to the ivorybill hunt in Arkansas has as its banner a photo of a flying Ivory-billed Woodpecker silhouetted directly overhead that was taken by the original Allen expedition to the Singer Tract.

"You're sure it was an ivorybill?" I asked, playing the skeptic. "What about an Anhinga? They have a long tail and might fool you if you saw them at an angle."

"No way, Geoff. This bird flew right over me. It was a big woodpecker, but it flew like a loon—straight and fast. I'm sure it was an Ivory-billed Woodpecker."

"I thought you said that you saw two."

"Yeah, well, the bird flew over, and I leapt out of my boat and started running through the woods after it. Tanner said in his book that he was able to chase down ivorybills this way. I ran into the woods for about 10 minutes without hearing or seeing anything, and then the mosquitoes got unbearable. I was in a huge cloud of them, and I couldn't take it. I ran back to my boat. I had just gotten back in my seat in the kayak when two Ivory-billed Woodpeckers flew over me from the direction I just ran. And this time I could see white flashing on the back side of their wings."

"So a bird flew over. You chased it, gave up, went back to your boat, and then two ivorybills flew back over the creek from where the first bird had gone?" I asked, making sure I had the sightings straight.

"Yep."

"Wow. What was their orientation? Did they fly side by side?" I knew from old accounts of ivorybills that when pairs traveled together, one bird flew a little ways behind the other bird. I wanted to see what Brian said about the positions of the two birds before I mentioned what I expected.

"No," Brian corrected. "One bird followed the other maybe 1 or 2 seconds behind. They went right over my head again. The shapes were perfect. And I could see white flashing on the back of their wings."

"Two birds," I thought. "Man oh man." Given the number of cavity trees and feeding signs that we had found in so many places on both sides of the river, I had been convinced that we were dealing with not one bird but at least a pair of ivorybills. Tyler kept reminding me, however, that all of our detections potentially fell within the home range of a single bird based on the home ranges of ivorybills in the Singer Tract. But now we had a sighting of two birds flying one behind the other just as is described in the ivorybill literature.

Having two ivorybills in this area was huge because it meant that we were likely dealing with a pair and that in the spring they should be nesting. Finding and documenting a nest was the ultimate dream of any ivorybill hunter. With two birds I thought we should also expect more vocalizations and double knocks in the winter, which meant much better chances to detect, observe, and get an image of the birds. As far as I knew, Brian was the first person to report a pair of ivorybills from Florida since Arthur Allen had watched a pair at a nest in 1924. With Tyler and Brian's sightings, I was now convinced that we really did have ivorybills in the forest around the mouth of Bruce Creek. The question was what to do next.

My Quandary 4

I'm not saying that I was in a position like George W. deciding whether to invade Iraq, but in the summer of 2005 I was faced with what seemed to me a big decision. Brian, Tyler, and I had found what I was now convinced were at least two Ivory-billed Woodpeckers in the Choctawhatchee River bottomlands in a remote part of the Florida Panhandle. Brian and Tyler had seen ivorybills and heard a kent call, and I had heard a double knock. We had found many large cavities and trees with scaled bark. The dilemma before me was what to do next.

I was in a unique situation. I am, after all, not simply a bird hobbiest or a kayaker who had stumbled across an ivorybill. I am a professional ornithologist, a full professor. Most of my research focused on the evolution of plumage coloration in songbirds, but I have published a dozen papers on avian conservation in the Southeast. The question was whether to hand our ivorybill discovery off to the already well-funded and organized Ivory-billed Woodpecker search team from the Cornell Laboratory of Ornithology, or to initiate our own study. This decision fell to me. Brian and Tyler were both in their early twenties and were just starting to think about becoming academic ornithologists. They had no experience with funding agencies or the politics of science. On the other hand, I had already waged many campaigns in the academic wars. I thought I had a pretty good idea where we stood.

The announcement a few months earlier by John Fitzpatrick and the Laboratory of Ornithology that they had proof of an Ivory-billed Wood-

Jerry Jackson in the Big Woods of Arkansas checking out 2004 ivorybill sightings for himself. (Photograph by Jerome A. Jackson, Jr.)

pecker in Arkansas had been greeted by the birdwatching community with rejoicing. It was a second chance for one of the most striking and elusive bird species in the world. The video that was the centerpiece of Fitzpatrick's announcement showed a tiny, blurry, fleeting image of a black and white bird, but in the article he showed point by point why it was, indisputably, an ivorybill. After the announcement of the discovery, everyone—amateurs and professionals alike—awaited follow-up announcements on how many Ivory-billed Woodpeckers constituted this population. The video left a lot to be desired, I thought, but now that an ivorybill was found, it should only be a matter of time before the search team obtained hours of professional-quality video.

By late summer 2005 when I was pondering my situation, the search effort in Arkansas had stalled. The web page hung in cyberspace unchanged since the day of the announcement. The much-anticipated updates never came. Two breeding seasons had been spent by a huge team of professionals and volunteers with little further evidence. The voices of skeptics were beginning to rival the cheers of supporters. As a matter of fact, several months later as I write this chapter, Jerry Jackson, an ornithology professor at Florida Gulf Coast University, whose distinguished research career focused on the study of North American woodpeckers and who is widely recognized as the leading authority on Ivory-billed Woodpeckers since the death of James Tanner, took Fitzpatrick and the Laboratory of Ornithology to task in an article in the *The Auk*, the top bird journal in the world. Jackson harshly criticized Fitzpatrick for what he characterized as a public relations-heavy, data-poor approach to documenting ivorybills in Arkansas. "Faith-based ornithology" was how Jackson described the Cornell search effort in Arkansas.

I could feel the weight of skepticism on Fitzpatrick and his colleagues, which was why I was confident that I could get the Cornell group interested in our discovery in Florida. I was hoping that maybe I could convince them to fund a follow-up search, perhaps even providing summer salary for me and technician positions for Brian and Tyler. In exchange I could put the Laboratory of Ornithology on an Ivory-billed Woodpecker population where they could channel their energy and money and escape the embarrassment of not being able to relocate ivorybills in Arkansas.

But the more I played out the scenario of calling up Fitzpatrick and handing him our Ivory-billed Woodpecker population, the less I liked it. I was concerned that the Laboratory of Ornithology, with its enormous resources, would take command, reducing us to secondary players in our own study. I had no doubt that Fitzpatrick and his colleagues would put the interests of the Ivory-billed Woodpecker first and do a professional job with the search, but I was worried that they would see conservation of the woodpecker best achieved if the center of operations was in upstate New York. I was enjoying the notion of an Ivory-billed Woodpecker search run from a university in the Deep South within the historic range of the species.

The idea that I was starting to like and that was strongly reinforced by the opinions of Tyler and Brian was to keep the discovery secret while we conducted our own small-scale follow-up search for the bird. We would use a small search crew that would camp in the swamp forest. We would live with the birds as did Arthur Allen, Paul Kellogg, George Sutton, and James Tanner in Louisiana when ivorybills were studied in the Singer Tract in the

1930s, drawing as little attention to ourselves among local people as possible. We would be a few researchers from Auburn University studying river-bottom birds. The forested tracts along the Choctawhatchee River were already owned almost entirely by the State of Florida, controlled by the Northwest Florida Water Management District, so we needed no permission to enter the land for wildlife observation. With this approach we would either confirm our sightings and begin a long-term study of the Ivory-billed Woodpeckers of the Florida Panhandle, or we would quietly fail to find the birds and with no fanfare enter our 2005 sightings into a Fish and Wildlife Service database. At least that was the idea under which we initiated our search. From afar, such a search always seems straightforward, and it is easy to suppose that the result will be either definitive proof or a failed effort.

In July, Tyler, Brian, and I started to formulate a plan. Brian was getting engrossed in the ivorybill hunt and was pursuing his growing passion for ivorybills with the same focused enthusiasm he had shown in the bluebird study. By July, Brian had decided to commit the next year of his life to search full time for ivorybills in the Choctawhatchee River wetlands. He would start as a technician (although I had no money for salary when he made the decision) and then become my graduate student, conducting a study of Ivory-billed Woodpeckers for his master's degree if our search succeeded.

It would have been great if Tyler could have joined Brian in the swamp for a full-time four-month search, but Tyler's goals were taking him in other directions. From the first season that Tyler worked for me, it was clear that he belonged in graduate school. Unfortunately, he hadn't even finished high school. We joked about him joining my lab as a master's student anyway, and I seriously would have taken him if we could have found a way to get him admitted without a bachelor's degree. (He had by this point received his GED.) But grad school without a bachelor's degree was a fantasy. Tyler disliked the idea of four years of undergraduate studies almost as much as he savored the idea of running his own research projects in grad school. In the winter of 2005 he had finally accepted the reality that he could only get the career and life that he wanted by going to college and getting his bachelor's degree. With a bachelor's degree he could get into graduate school, get his Ph.D. in ornithology, and become a professor. He was scheduled to start his undergrad program at a regional college in Kansas in the fall of 2005, which meant that he would not be available to search the Choctawhatchee

Dan Mennill paddling Bruce Creek. Most of Dan's ivorybill hunting was done in front of a computer screen searching sound files, and the payoff was numerous kent calls and double knocks. (Photograph by Geoffrey E. Hill.)

River basin for ivorybills full time in the winter and spring of 2006. It was a hard irony that after half a decade of total freedom, he was tightly scheduled for the spring of our ivorybill search. The thought of having to participate in the ivorybill study part time was agonizing for Tyler, but to his credit he stuck with his plans for college. He would still have a full month during the December/January semester break and a week during spring break to be in the swamp searching for ivorybills with Brian.

With Tyler planning to be on site for a month around Christmas and Brian committed to a full-time six-month search, things were falling into place, but I felt that our search really needed one additional tool—remote sound recorders. Remote sound recording devices could be set up in areas where Ivory-billed Woodpeckers were suspected to be present and potentially record audio evidence of the birds. I thought that such recorders might give us not only tangible evidence for the existence of ivorybills but also guide us in our search for roost or nest cavities. The problem was that the Laboratory of Ornithology had invented such stations, which they called autonomous recording units (ARUs). Virtually everyone in the world who knew how to construct and run remote sound monitoring stations was at the Laboratory of Ornithology, and we couldn't go to them for equipment without revealing our ivorybill study. Besides, word on the street was that Cornell's stations cost about $8000 per unit, and at this point we had zero budget. By an enormous stroke of good fortune, Dan Mennill, one of the few people outside of Cornell who was an expert in the construction and use of remote sound monitoring stations, had an office two doors down the hall from mine at Auburn and was my postdoc at the time.

Dan had been a fortuitous recruit to my lab group. His wife, Stephanie Doucet, wanted to study plumage coloration for her dissertation and had sought me out as a potential graduate advisor when she completed her master's degree at Queen's University in Ontario. My primary focus as an ornithology professor (my day job, so to speak) is studying the function and evolution of ornamental coloration in the feathers of birds. Most of my published work has focused on whether the variably red-to-yellow plumage coloration of House Finches is used by females in choosing mates and whether the ornamental coloration might encode information about the quality of the male. The bluebird study that I mentioned earlier was a relatively recent excursion into the function of structural coloration: bluebirds don't get their blue coloration from blue pigments; the microstructure of the feather interacts with light to create the color display. During the summer of 2005 I was consumed by duties as editor for a huge two-volume compendium entitled *Bird Coloration.* Any ornithologist who might know my work would know me as an expert in plumage coloration, certainly not as an expert on woodpeckers or conservation biology. Ivorybill hunting was just a hobby I took up in May 2005 while I was waiting for page proofs for my book.

From the outset, Stephanie had made it clear that she and Dan were a package deal. They are perhaps the closest young couple I've ever met, sharing all aspects of both their work and private lives. They would not consider being apart for an extended period of time; to recruit Stephanie I also had to recruit Dan. Dan was due to receive his Ph.D. just a few months before Stephanie would start graduate studies with me, and he was looking for a place to be a postdoc—a short-term position taken by scientists who have received their Ph.D. but have not yet earned a permanent university position. Stephanie and Dan made a very enticing package. Stephanie had already published three papers in top journals, a very good rate of publication for a master's student, and had just won the student paper competition at the American Ornithologists' Union meeting (coincidentally presided over by Fitzpatrick, then president of the American Ornithologists' Union).

Dan was not only a very talented young behavioral ecologist, having just won the Allee Award, a sort of international rookie-of-the-year award in animal behavior, he also had his own salary in the form of a Natural Sciences and Engineering Research Council (NSERC, Canadian government) postdoctoral fellowship. All he needed was an office and a decent working environment. I convinced the pair that Dan would be well taken care of at Auburn, and I pulled off the greatest recruiting coup of my life. Stephanie and Dan moved to Auburn in 2002. Stephanie was a fabulously productive student, developing an exciting project on color displays in Long-tailed Manakins in Costa Rica, and Dan was a tremendous asset to my lab group, working with me on song and mate choice in House Finches and serving as a statistical consultant and a sounding board for all of my grad students. If Dan would help us with the woodpecker search, then my recruiting coup had the potential to pay off in an unanticipated way.

Tyler, Brian, and I approached Dan and told him that we wanted to discuss an idea for a research project, but that we had to have his word that what we were going to discuss would remain confidential. Dan was understandably a bit puzzled. All good research ideas are valued by scientists. Members of my group discuss ideas with an understanding that we keep each other's ideas in confidence. Obviously, what we had to discuss was something extraordinary, or we wouldn't demand a promise of confidentiality up front. Dan agreed to our confidentiality request, and we told him our tale of the Ivory-billed Woodpecker discovery, our idea for a follow-up search, and the need for sound monitoring in the area. With no hesitation Dan enthusiastically agreed to join the search team and be in charge of sound monitoring. The situation was complicated a bit by the fact that Dan had accepted a faculty position at the University of Windsor in Ontario and would move to Windsor in August. However, with his start-up money at Windsor, he intended to buy equipment that could be used to construct sound monitoring stations. The new sound equipment would be for his ongoing studies of tropical wrens in Costa Rica, but he wouldn't need the equipment in Costa Rica until May, so until then we could use it in Florida to try to record ivorybills. Long before we invited Dan to join our Ivory-billed Woodpecker search, he had dreamed of establishing a center for the study of animal sounds when he got a faculty position in Canada. Establishing his lab in Windsor as the center for the study of Ivory-billed Woodpecker acoustics would provide a great platform from which to build his animal-sounds laboratory.

So we had lined up our search crew—Brian permanently on site from January to May, me coming down on weekends as often as possible, Tyler on site over Christmas break and spring break, Dan in Canada analyzing the recordings that were sent back from the swamp. Dan would add one more key member to the crew when he persuaded his newly recruited master's student, Kyle Swiston, to join Brian full time in the swamp from January to May. I had been worried about leaving Brian alone in this remote swamp location for weeks at a time. Now he would have a companion. Kyle would run the listening stations and would be in charge of on-site analysis of the sound recordings. Whether we relocated Ivory-billed Woodpeckers or not, Kyle could use the recordings made at listening stations as the basis for his master's degree at Windsor. As a matter of fact, Kyle wrote a nice research proposal for the study of woodpecker communities on the Florida Panhandle with no mention of the Ivory-billed Woodpecker.

The remaining problem was funding. Even though we were planning to conduct this project on a minimal budget, it could not be done out of pocket—at least not the shallow pockets of a professor. To stay financially solvent during the search Brian needed a small stipend, about $1,000 per month. We also needed some basic gear and a generator so that we could recharge the batteries for our video cameras and listening stations at our remote field camp. We thought we could get a bare-bones search done for about $10,000. I imagine that this was approximately the *daily* budget for the massive Laboratory of Ornithology search in Arkansas, but it was not a trivial sum of money for us. The problem was that we had far too little evidence at that point to go to a national science or wildlife funding agency for support (and there would be no way to keep our search secret if we did apply for such funding).

In August 2005, I made an appointment to see Stewart Schneller, dean of the College of Science and Mathematics at Auburn University and my boss. I brought with me the *Science* article about the Arkansas discovery with the Ivory-billed Woodpecker on the cover, maps of the Choctawhatchee River basin, and a research budget proposal for $10,000. Auburn University is a state agricultural school in a poor southern state. It does not have the endowments of rich private schools, and the various science programs at Auburn have been only moderately successful over the past decade at bringing in the sort of large federal grants that would give administrators overhead dollars to use as discretionary funds. I knew that $10,000 was a lot to ask my dean to come up with, especially when I had no definitive evidence that we had actually discovered Ivory-billed Woodpeckers.

Stewart is a chemist, but he had always been very supportive of my bird research, and we have a good working relationship. He listened as I explained our discovery and my plans for a follow-up search. I told him that I thought our ivorybill discovery could make Auburn a center for conservation biology in the Southeast, but my spinning was unnecessary. Stewart did not need my help to connect the dots. He immediately saw that this could be very big for Auburn. He glanced at the budget and said he would be happy to award me a discretionary grant of $10,000 to undertake our search.

I left the office feeling very fortunate that I was at an institution that was willing to back such a search, but also more nervous than ever about the coming field season. The stakes were getting higher and higher. It was up to our little search crew now. We knew where the birds had been in the spring and summer of 2005. We had the equipment and personnel to document them. Would we be able to relocate and document Ivory-billed Woodpeckers during the breeding season of 2006? All it would take would be a bit of expert field ornithology and some luck. I felt like we were on a roll, and I was hoping it would continue.

Is It a Miracle?   5

How did the most sensational, long-vanished bird in North America suddenly materialize in the twenty-first century in one of the most densely populated states? The persistence of a population of Ivory-billed Woodpeckers in the Sunshine State seems like wild fiction. But the real fiction was the claim by some that there was no place where a large forest bird could go undetected in Florida. Anyone who thinks that Florida is one big go-cart park, suburb, and beach resort has a basic misconception of the state.

As innumerable writers have commented, Florida is a state of contrasts. It is the home of countless retired Yankees who flee the depressing gray and stiffening cold of northern winters. It is the playground of the rich and famous, the middle-class family, and college spring breakers drawn like flocks of Sanderlings to the beautiful clear water and white sand beaches. It is the relocation site for untold thousands of young families who fill suburbs that sprawl over former citrus orchards and wetlands. It is Disney World and Cape Canaveral, South Beach and the Daytona Speedway. But this familiar, crowded, and overdeveloped face of Florida is found mostly in the central and southern parts of the state. Those regions left Dixie long ago, as the central and southern Florida Peninsula became a transplanted appendage of the urban north. South of Gainesville, you can barely find a Baptist Church, and you have to ask to get sugar in your iced tea.

Things are different in the Panhandle, especially west of Tallahassee. Here people have a population density and a culture shared with southern

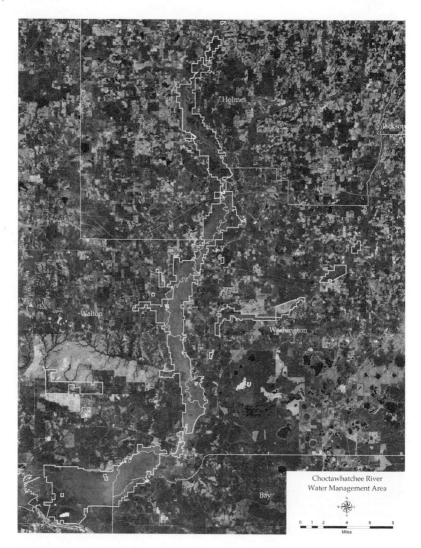

Choctawhatchee River
Water Management Area

0 1 2   4   6   8
            Miles

A satellite photo of the Choc-
tawhatchee River basin showing
boundaries of land owned by
the Northwest Florida Water
Management District.

Alabama and Georgia. Except for a few growing beach re-
sorts, there are no sprawling suburbs, and the landscape is
largely forested. The Florida Panhandle is primarily a rural
community where folks typically live simple lives that revolve around family, the local church, and hunting and fishing. They rarely have college ed-
ucations, mostly derive their incomes from the land (farming or cutting
trees), and as much as is possible in a satellite-TV world, live lives isolated

from much of the rest of the country. If an Ivory-billed Woodpecker excavated a cavity in the yard of most of the people living in the rural Florida Panhandle and if they happened to take note of it, they would dismiss it as just a "peckerwood," hardly something worth getting worked up about. I think it would be rare for anyone living near a swamp in rural northern Florida to recognize an Ivory-billed Woodpecker as something noteworthy, and it would be even rarer if they reported a sighting to an authority. Even if they told someone, probably the sheriff or a local game and fish officer, the chances of that person passing the information along to appropriate ears are slim. And if by chance their observation made its way to a regional bird club, Audubon society, or wildlife department at a university, it is unlikely that an uneducated local who did not know the correct name for the bird or how to properly describe field marks would be taken seriously.

All this is to say that people have been living among and seeing Ivory-billed Woodpeckers in the Florida Panhandle since Europeans first settled in the area a couple of hundred years ago—Native Americans, of course, knew the bird well. Local fisher-

An aerial photo of the forest corridor along the Choctawhatchee River near Lost Lake.

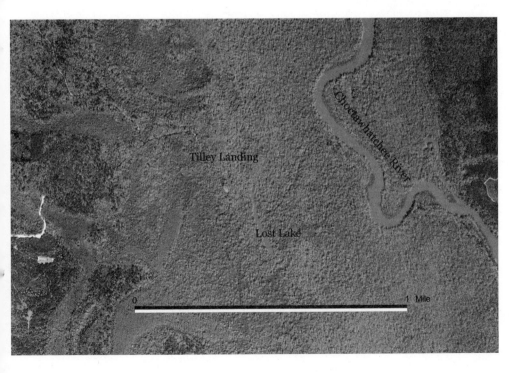

men and hunters undoubtedly see Ivory-billed Woodpeckers at least occasionally as they venture into the forests along the Choctawhatchee River, and some of these outdoorsmen may even recognize the bird as somethiing special. Unfortunately, the gulf between the world of the rural southerner and the world of the professional ornithologist is too wide to allow for exchange of information. A hundred years ago collectors served as conduits for information flow, enlisting local people to collect birds for them and selling them to museum men and wealthy collectors. Now, all such links are broken. The often-stated assertion of the ivorybill skeptic that if Ivory-billed Woodpeckers were there, people would see them, is correct. The flaw in the logic is the assumption that such sightings would make their way to authorities, be taken seriously, and lead to the discovery of the birds by experts.

The Choctawhatchee River and the swamp forests of the Florida Panhandle have several attributes that make them likely Ivory-billed Woodpecker refuges. First, the swamp forest along the Choctawhatchee River is big. This seasonally flooded forest extends from the town of Geneva, Alabama, to Choctawhatchee Bay, Florida, along about 60 miles of river. South of Interstate 10 on the Florida Panhandle the flood forest along the Choctawhatchee has an average width of about 1.5 miles, ranging from less than 0.5 miles in a few pinch points to more than 3 miles in the lower delta. This broad, continuous forest runs for 40 miles to Choctawhatchee Bay, and it lies in a mostly forested landscape. With few exceptions, this swamp is roadless. From Interstate 10 the Choctawhatchee River flows south for about 25 miles with no road crossing it until Highway 20 at Bruce. The river then completes its final 15 miles to Choctawhatchee Bay with no more road crossings. That's one road crossing 40 miles of swamp forest.

A second important feature of the Choctawhatchee River bottomlands is that much of this forest was only selectively logged. According to some local residents who know the history of the river, commercial logging in the wet forests along the lower Choctawhatchee River occurred primarily in the 1920s and 1930s. This coincides well with what James Tanner reported in his book *The Ivory-billed Woodpecker* as the peak period of commercial logging in the cypress stands of the northern Gulf Coast. Logging in the forests along the Choctawhatchee was rapid, selective, and sloppy. Rail lines were run into the forest (we can see the ruins of an old logging bridge near the mouth of Bruce Creek). Large cypresses with straight, unblemished trunks were targeted. Huge (usually hollow), imperfect, and younger cypress trees were left standing. The harvest was very patchy, with hardwoods as well as cypresses targeted in some areas, but with large stands of oak, tupelo, and sweetgum left untouched in other areas. Sites that were easily accessible near water channels or rail lines tended to be intensively harvested, and some hard-to-reach spots were skipped entirely.

This account of the logging of the area coincides well with what we see in the current forest. There are huge cypresses, almost always misshapen and hollow, scattered throughout essentially all wet areas of the basin. Along with these scattered giants, there are many stands of large cypress trees, probably about 100 years old, that would have been small trees 80 years ago. But there are relatively few cypress trees between these size classes that would have been in their prime during the harvest in the early twentieth century. Most of the logging in the swamp forests in this basin was done before the chain saw was invented. Trees were cut one at a time with handsaws

Mark Liu paddling next to the ruins of a logging bridge near the mouth of Bruce Creek. (Photograph by Geoffrey E. Hill.)

and axes. No wonder the lumberman only took the prime timber and left many of the largest, but imperfect, cypresses alone.

There is no reason to assume that this sort of lumbering operation would have driven the Ivory-billed Woodpecker to extinction in this river basin. Undoubtedly, birds would have fled the noise and human presence in the sections of the forest that were being actively logged. But timber extraction would have been going on in only a small portion of this basin at any one time. Because logging was selective, there was still a forest standing when the lumbermen moved on, and one could imagine that Ivory-billed Woodpeckers even sought out the recently disturbed stands of forest if the logging activity left dying trees. Because many large trees were left throughout the basin, the forest was able to recover.

Two colleagues, Mark Bailey and Matt Aresco, who are experts at mapping and in whom I confided about our discovery, found aerial photos of the Choctwhatchee River near Bruce Creek taken in 1941, just after the most intensive logging of the wetlands would have been completed. The bottomlands were completely forested in 1941. There are no large blocks of clearcut forest. These maps show that timber extraction along the Choctawhatchee River was selective, and there was never a time when the bottomlands were deforested. As a matter of fact, over the entire landscape, there is probably less forest today than in 1941 because large areas of upland forest have now been cleared for pastureland and housing or converted to short-rotation slash pine plantation. The swamp forest, however, appears to have completely recovered and stands today as a mature forest. Structurally, large areas of the bottomland forest along the Choctawhatchee River are, in my opinion, nearly identical to a virgin bottomland forest.

That the forests we were searching along the Choctawhatchee River really are fully mature forests was made clear to me in November 2005. My girlfriend, Wendy Hood, lives in Myrtle Beach, South Carolina, where she's a professor at Coastal Carolina University. We visit as often as our busy schedules allow, and it's an interesting footnote to this whole adventure that if Wendy hadn't gone on a month-long trip to Africa in May 2005, I would likely have been visiting her instead of floating the Pea River, and the ivorybills of the Florida Panhandle would undoubtedly still be undiscovered. Anyway, one weekend in November 2005 we decided that instead of me driving into Myrtle Beach, we would meet in Columbia and visit Congaree Swamp National Park. The Congaree Swamp lies on the north side of the Congaree River just above the point where the Congaree River joins the

Wateree River to form the Santee River. It is touted by the National Park Service on the park's brochure as "the largest intact tract of old-growth floodplain forest in North America." So I took Congaree Swamp as the gold standard of swamp forests in the southeastern United States. Wendy and I spent a leisurely day paddling through the heart of the swamp on Cedar Creek. It was a beautiful old forest with huge trees, but it was not completely virgin timber—a detail that I found especially irritating given a conversation I had had earlier in the day.

At the information center before we started, I asked the ranger at the help desk where we would find the most impressive virgin forest. I tried to make it clear that we wanted to see what was considered to be untouched swamp forest.

"Son," he said in an extremely condescending voice as he peered over the top of his glasses, "it's all virgin."

"Really?" I answered, ignoring the condescending tone. "If we paddle down Cedar Creek we'll be in the best old-growth forest in the park?"

He looked so exasperated he seemed barely able to form an answer. "That's as nice as any area. Like I said, it's all uncut timber."

Ten minutes into our float down Cedar Creek I was staring at old cypress stumps that had clearly been cut with a saw. There weren't many stumps—the impact of humans had been minimal in this forest—but still, as I reflected on the arrogant tone of the park ranger, I thought, "What a jackass. That guy really should get out in a kayak with his eyes open."

But despite the fact that a few cypresses had been cut along the creek, it was a magnificent forest. Enormous oaks and cypresses draped in Spanish moss towered over the creek. As we drifted along the placid creek beneath the huge trees, it gave me the feeling of being in a primordial swamp. It also gave me a strong sense of déjà vu. Cedar Creek in Congaree Swamp was the twin to Bruce Creek in Choctawhatchee River Swamp. I'm happy to report that the Congaree Swamp is not "the largest intact tract of old-growth floodplain forest in North America." It is a very distant second to the Choctawhatchee River bottomlands. At the time of this writing I have yet to tour the Apalachicola and Escambia rivers, among other rivers on the northern Gulf Coast, and these may come in ahead of Congaree as larger intact tracts of old-growth forest. By the end of my paddle through the Congaree Swamp, which is a very special place, I was getting the feeling the Choctawhatchee River bottomlands were equally special. And I should mention that I saw no signs of ivorybills—no feeding trees and no sugges-

(Left) Wendy Hood paddles Cedar Creek in Congaree National Park. (Right) Bruce Creek near the Choctawhatchee River. Both creeks run through beautiful old-growth bottomland forests. (Photographs by Geoffrey E. Hill.)

tive cavities—anywhere in the Congaree, but my one-day paddle was hardly an exhaustive search.

Choctawhatchee River bottomlands are not only extensive and mature, but they lie in a forested landscape. Unlike the fertile lowlands around the Mississippi River that make some of the most productive cropland on the continent, the sandy soils of the Florida Panhandle are poor for row crops and even for pastures. Back in the 1930s, when Tanner was studying Ivory-billed Woodpecker in the Singer Tract in Louisiana, these birds were already living in an island of habitat surrounded on all sides by fields of wheat, corn, and soybeans. Greenlea Bend, the area that Tanner singled out as the gem of the entire Singer Tract, was largely deforested soon after Tanner's study and is now mostly a soybean field. In the Florida Panhandle, in contrast, seasonally flooded bottomland forests are rarely converted to agricultural fields. If a bottomland forest is cut, it is usually left to regrow as a forest. Even areas out of the seasonally flooded bottomlands are mostly managed

as pine plantations. In the same way that the Mississippi River bottomlands are a center for agriculture, the northern Gulf Coast has become a center of the paper-pulp industry. A vast acreage of the uplands in this area is a managed pine forest. The native stands of longleaf pine have almost completely given way to rows of slash or loblolly pine. Conversion of longleaf pine savannah to commercial tree farms was an ecological disaster; for many species it was as devastating as conversion of forests to row crops. From the perspective of an Ivory-billed Woodpecker, however, having swamp forest abut a sea of commercial pine forest is much better than having the swamp border a sea of bare soil. Even if Ivory-billed Woodpeckers don't forage in commercial pine stands (and my inspection of slash pine groves adjacent to swamp forest suggests they may), they can disperse through pine forests between river systems and isolated bays much more easily than they can fly across open agricultural fields.

Another feature that makes Choctawhatchee River a plausible refuge for

Ivory-billed Woodpeckers and that distinguishes the Choctawhatchee River bottoms from the swamp forests of Louisiana and Arkansas is that the swamp forests of the Florida Panhandle receive much less hunting pressure. This statement might seem like an oxymoron given that hunting is second only to religion in the lives of most residents of the Florida Panhandle. A drive around Washington or Walton counties, where our study site is located, on any day in November or December will convince you that every male over the age of nine is out with a gun in his hand. But hunting occurs mostly in the upland forests. Compared to Louisiana or Arkansas, almost no ducks winter in flooded forests in the Florida Panhandle—only Wood Ducks winter in any numbers. Consequently, there is little duck hunting and only by a few local people. The flooded forests of Arkansas and Louisiana, in contrast, are the dream travel destinations for every waterfowl hunter in North America, and the swamp forests of eastern Arkansas host thousands of hunters each fall. Deer and turkey are hunted intensively in the uplands adjacent to the Choctawhatchee River bottomlands but rarely in the swamp itself, especially when the swamp is flooded.

With limited game to pursue in forests, people mostly enter the area to fish. Fishing pressure is relatively heavy along the Choctawhatchee River, and in April and May, during the best season for fishing, there are typically small fishing boats every half mile or so along the river. Fishermen stick to the main river and large creeks and channels and do not penetrate very far into the swamp. I can't see how fishermen, sitting quietly in their boats, would bother ivorybills. They may regularly see ivorybills, but they wouldn't harm or greatly disturb them. We find the roar of boat motors irritating when they break the solitude of the swamp, but I imagine that ivorybills, which would have heard such motor noise their entire lives, don't pay much attention to it.

Because fishermen stick to main channels and because few hunters penetrate deep into the flooded forests, there are large areas of swamp forest along the Choctawhatchee River that I think are never visited by people, and no area of the swamp forest is known to more than a handful of local people. When I walked through the forest south of the mouth of Bruce Creek in November 2005, I saw no signs of humans. I now know that a local hunter named Earl regularly visits this area, but he leaves no sign of his presence. There were no shell casings, no flagging, not a single piece of litter. Similarly, when I later paddled through large sections of flooded forest south of Lost Lake and across the river from Bruce Creek, I covered miles of flooded forest that bore not a single sign of people save the 100-year-old saw marks on the cypress stumps. Where people go, they tend to leave a lot of litter (beer cans, wrappers, fishing garbage), so lack of garbage in an area is a good indication that people are absent. Anyone who thinks there is no place left in the eastern United States that does not feel the constant pressure of humans should spend some time in these swamp forests. There is plenty of space for Ivory-billed Woodpeckers to live undetected in the Choctawhatchee River swamps.

Based on habitat and location, it's not ridiculous to suggest that Ivory-billed Woodpeckers persisted in this swamp. But that doesn't explain how ivorybills went undiscovered for so long. Why weren't they found sooner? As far as I can determine, no one with book knowledge of birds has been in the swamps of the Choctawhatchee River since the ivorybill was declared extinct in the middle of the twentieth century. The claim that the Choctawhatchee River is unbirded seems amazing given that it is the third largest river in Florida, but consider the *Florida Breeding Bird Atlas*, completed in 1992. This atlas project was designed to determine the distributions of all breeding species in Florida. Some of the gaps in breeding records for the Choctawhatchee River basin suggest that no one spent time in the swamp forests along this river. For instance, the Swallow-tailed Kite is listed only as a possible breeder along the Choctawhatchee. In the spring of 2006, however, we found Swallow-tailed Kites to be a common, conspicuous, and widespread breeding bird throughout the river bottomlands. With no effort we observed adults carrying sticks as evidence of nesting, and adults flying with young later in the spring. There are no records at all of Yellow-crowned Night-Herons in the Choctawhatchee River basin, even though this is an abundant bird whose numerous nests along creeks simply could not be missed by someone conducting a bird census deep in the swamp. Even the Hairy Woodpecker, which is a common breeding bird, is listed only as a possible breeder. It seems to me that no one ever got in a boat and penetrated into the swamps along the Choctawhatchee to gather data for the atlas, and the breeding bird atlas was the best reason for local birders to conduct an ornithological investigation in this area.

Even though birders never got into this swamp, ivorybill hunters should seek out remote swamps. How could the forests along the Choctawhatchee River have remained so far off of the radar screens of ivorybill searchers for so long? The failure to recognize the Choctawhatchee River bottomlands as a potential area for Ivory-billed Woodpeckers begins in the late nineteenth century when collectors were systematically exterminating the remaining Ivory-billed Woodpecker populations in Florida. The accounts in Tanner's book and in Jerry Jackson's *In Search of the Ivory-billed Woodpecker* are heartrending. Just when the populations of Ivory-billed Woodpeckers reached their lowest points after the wholesale cutting of southern swamps in the late nineteenth and early twentieth centuries, collectors went after the remaining birds, systematically shooting every Ivory-billed Woodpecker from all known localities. This slaughter was not just accepted but actively encouraged by the leading American ornithologists of the day. Prominent ornithologists such as Frank Chapman, curator of birds at the American Museum of Natural History and originator of the Christmas Bird Count, and William Brewster personally shot ivorybills when they knew there were few individuals remaining, and they paid for the skins of ivorybills shot by professional collectors, encouraging them to seek out and kill the last birds.

Habitat destruction is invariably cited as the reason for the disappearance of ivorybills in North America, but I'm convinced this conclusion is only partly correct. No doubt, the wholesale cutting of the great cypress swamps of the South enormously reduced and fragmented ivorybill populations, just as the felling of the great longleaf pine forests of the Southeast reduced and fragmented populations of Red-cockaded Woodpeckers. If people had just cut the cypress and ignored the birds, however, I feel confident that scattered populations of ivorybills would exist across the South today just as Red-cockaded Woodpeckers survived in a few populations. It was the slaughter by collectors, systematically shooting their way through these small, vulnerable populations in isolated patches of forest, that pushed the ivorybill to extinction (or, as I'm happy to report in this book, close to extinction).

Locales where ivorybills were collected in the late nineteenth and early twentieth centuries became the primary basis for delimiting the range of the species. Tanner first compiled collecting locales from museum specimens and plotted them on maps of the Southeast and the Mississippi River Valley. Jackson refined these maps with additional collection and sight records when he wrote his book and a *Birds of North America* species account on ivorybills sixty years later. Any Ivory-billed Woodpecker enthusiast will know that neither Tanner nor Jackson found any records for the Choctawhatchee River. There are few or no old sight records, and no known specimens of Ivory-billed Woodpeckers from any of the rivers between the Alabama River above Mobile, Alabama, and the Apalachicola

William Brewster, one of the leading American ornithologists at the turn of the twentieth century, who collected an Ivory-billed Woodpecker along the Suwannee River in Florida. (Courtesy of the Ernst Mayr Library of the Museum of Comparative Zoology, Harvard University.)

River to the east of the Choctawhatchee River, except one bird collected along the Conecuh River in Alabama. Why did collectors apparently miss the ivorybills in the Choctawhatchee River swamps? I have no good answers. Maybe Choctawhatchee ivorybills were missed by collectors for the same reasons that one nice stand of cypress in an entire region of a state gets missed by lumbermen: chance, oversight, inconvenient location. Maybe for some reason the ivorybills that lived along the Choctawhatchee River were warier and harder to collect than ivorybills in other areas. There is no way to know. Let's just be glad that Choctawhatchee ivorybills were passed over, or I imagine that collectors would have exterminated Ivory-billed Woodpeckers along the Choctawhatchee River as efficiently as they exterminated the species in other parts of its range.

Lack of specimens and no written accounts led Tanner to exclude the Choctawhatchee River bottomlands from consideration as a search area. Tanner was the first ornithologist to attempt to identify all the potential suitable areas for Ivory-billed Woodpeckers in the United States. I'm a bit reluctant to criticize his conclusions. It's sort of like criticizing Darwin, who laid the foundation for modern evolutionary biology. Actually that's not quite right; evolutionary biologists criticize the work of Darwin all the time. Criticizing Tanner is more like criticizing a patron saint if you're Catholic. Tanner looms as the patron saint of the Ivory-billed Woodpecker, and his dissertation on the natural history of the woodpeckers in the Singer Tract is revered like a bible by some ivorybill enthusiasts. Without a doubt, Tanner was a great field biologist. Few academics could have chased Ivory-billed Woodpeckers from their roost cavities for a mile through the swamp woods to their morning feeding trees. Few could have endured the daily rigors of a remote camp in a swamp woods that allowed him to record details of the life history of the birds. These were Tanner's strengths, and his natural history of the Ivory-billed Woodpecker in the last great forest in the lower Mississippi River Valley deserves the appreciation it gets.

In his book, Tanner wrote a chapter, "Present Distribution and Numbers of Ivory-billed Woodpeckers," in which he came to the conclusion that in 1939 (the date stated for the population estimates given in a table) there were approximately twenty-two individual Ivory-billed Woodpeckers left in a total of five locations in the United States. This conclusion was based on eight months of searching, visiting remaining large swamp areas, and talking to local woodsmen across the former range of the species. In my opinion, Tanner's attempt at a one-man inventory of all Ivory-billed Woodpeckers and ivorybill habitat in the United States was perhaps the greatest folly in the history of Ivory-billed Woodpecker conservation and one of the greatest follies in the history of U.S. bird conservation. Tanner wrote that no more than six birds persisted in any population and that the trees were being cut out from under all the remaining birds. His opinion on ivorybills deservedly carried great weight. If Tanner said that in 1939 we were twenty-two birds away from the end of the species, then extinction was inevitable. And once ivorybills had been pronounced extinct by the greatest ivorybill expert in the world, it became virtually impossible for the species to be resurrected. Only crackpots and kooks, after all, see extinct birds. And if the bird was gone, there was no good argument for preserving its habitat.

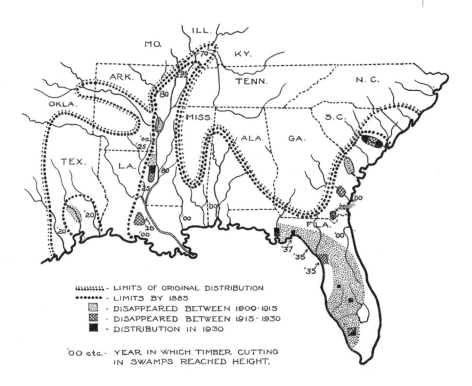

=== - LIMITS OF ORIGINAL DISTRIBUTION
•••••• - LIMITS BY 1885
▦ - DISAPPEARED BETWEEN 1900-1915
▧ - DISAPPEARED BETWEEN 1915-1930
■ - DISTRIBUTION IN 1930

'OO etc.- YEAR IN WHICH TIMBER CUTTING
IN SWAMPS REACHED HEIGHT.

A map by James Tanner from his book The Ivory-billed Woodpecker *showing the historic distribution of ivorybills in the southeastern United States, including the only areas where he thought they remained in 1930. (*The Ivory-billed Woodpecker. *Research Report Number 1, National Audubon Society, New York, 1943.)*

I find it inconceivable that one person would think that he could exhaustively inventory a large portion of a continent for a secretive animal that inhabits remote swamp forests. Think of how difficult it was for teams of biologists to inventory Cuba for ivorybills in the late twentieth century. The pine forests of Cuba constitute a miniscule area compared to the area covered by swamp forest in the southeastern United States. To understand why Tanner thought such an inventory was feasible, we have to consider the assumptions and methods on which his survey was based. First, Tanner held up the Singer Tract in Louisiana as the standard to which all other potential ivorybill areas must be compared. The Singer Tract was an 81,000-acre virgin swamp wilderness when Tanner conducted his ivorybill studies, and it supported only about six pairs of ivorybills—not enough birds for a viable population. Tanner acknowledged that the ivorybills in the Singer Tract might not have been at maximum capacity when he studied them and that birds might have lived at higher densities elsewhere, particularly in Florida, but he never got away from using the Singer Tract as the benchmark for the area needed to support a population of ivorybills. Basically, Tanner laid a Singer Tract–sized template across the remaining forested wetlands of the southeastern United States and he found them all wanting. There were no areas of virgin timber even close to the size of the Singer Tract remaining in 1939, so given the assumption that an area like the Singer Tract was the minimum needed for a viable population, he concluded that there couldn't be more than remnant doomed ivorybills remaining.

It was the idea that ivorybills have to have immense stands of completely virgin timber, plus the lack of historical records, that caused Tanner never to consider the Choctawhatchee River or several other likely river basins in the Florida Panhandle as Ivory-billed Woodpecker habitats. Subsequent ivorybill hunters, I think, put too much stock in Tanner's assessment, and in so doing missed several obvious places to look for the birds. Basically, once the suggestion that the birds were extinct took hold, no one really ever went back and challenged the assumptions on which the pronouncement was made.

I think the notion that ivorybills can only persist in virgin timber with 1,000-year-old trees is particularly flawed when one considers forests along the Gulf Coast, which historically were a center of abundance of ivorybills. Long before the vast tracts of cypresses growing in these Gulf forests felt the stinging axes and saws of lumbermen, they regularly endured the destructive force of hurricanes. If you pay attention as you paddle around the Choctawhatchee River bottomlands, you notice that the old cypresses are invariable flat topped, with few large and spreading branches. This is a tree morphology shaped by dozens of hurricanes over centuries. Virtually all the old trees have been broken, bent, and beaten by hurricanes. In our main study area, there is a large area of downed trees where a tornado ripped through the forest in the last couple of years, and this is the area in which we have had numerous encounters with ivorybills. Ivory-billed Woodpeckers seem to use these tornado-damaged trees as a primary feeding area. Hurricanes and the tornadoes they spin off are regular events along the entire Gulf and Atlantic Coast range of the ivorybill, and in this hurricane belt, no forest stands undisturbed for more than a few decades. If ivorybills required forests that grew untouched for a millennium, they would never have been found anywhere near the Gulf or southern Atlanta coasts.

So perhaps Tanner used a flawed method for pronouncing the extinction of the ivorybill, but how did all the searches conducted later in the twentieth century miss an entire population? My response to this is, "What searches?" Tanner and Jackson missed the Florida population because historical data led them elsewhere, but these were two men trying to search a continent. Who else really searched by getting deep into swamps for extended periods of time? Through the end of the twentieth century, virtually no one. John Dennis, an ivorybill hunter in the mid-twentieth century who seems to have been viewed somewhere between a respected professional ornithologist and a cryptozoologist, may have been one of the few individuals who carefully searched some of the swamp forests of the Southeast, doing most of his searching in the 1960s and 1970s. Perhaps not coincidentally, Dennis reported finding ivorybills along the Chipola River in Florida, in the Big Thicket region of Texas, as well as in Cuba. The fact that he claimed to have observed ivorybills is a primary reason he suffered from credibility problems—again, only kooks or crackpots see extinct birds. His Cuba sighting could not be dismissed because he took a photograph of a bird. Despite obtaining recordings of possible kent calls in Texas, Dennis was ridiculed for his sighting in the Big Thicket.

You might think Dennis's success in finding ivorybills would have spurred more widespread searches. To my knowledge it didn't. The idea that there have been massive, organized searches for Ivory-billed Woodpeckers covering every corner of every woodlot in North America is one of the great fallacies of the Ivory-billed Woodpecker legend. Tim Gallagher and Bobby Harrison were devoted ivorybill hunters in the 1990s and 2000s, and yet as Gallagher recounts in his book *The Grail Bird*, until their Cache River sighting they did all of their searching from roads and the edge of forests. Gallagher and Harrison had scarcely ever been in a canoe, let alone a kayak, when they took their famous trip down the Cache River in 2004. If Brian, Tyler, and I had tried to look for ivorybills along the Choctawhatchee River from roads, landings, and bridges, we'd have never gotten close to the birds. In the late twentieth century, the ivorybill was declared extinct by most birders and ornithologists with no concerted effort having been made to find the birds.

There have been more thorough searches for ivorybills since 2001 than in the entire second half of the twentieth century. This surge of interest in ivorybills is undeniably because the Laboratory of Ornithology, led by John Fitzpatrick, legitimized ivorybill hunting when it started to conduct organized and publicized searches. It began when a wildlife student at Louisiana State University, David Kulivan, reported seeing a pair of ivorybills while turkey hunting along the Pearl River in Louisiana in April 1999. Fitzpatrick and his lab, in collaboration with Van Remsen, an ornithology professor at Louisiana State University, organized a follow-up search. This was the first organized, publicized search for ivorybills in North America since Arthur Allen and his crew visited the Singer Tract in the 1930s, and it generated a lot of excitement about ivorybills.

Throughout this book I take a rather critical view of how Fitzpatrick and the Laboratory of Ornithology handled evidence that they gathered in Arkansas and particularly how they presented the Luneau video. It would be terribly unfair of me not to point out how much good the Laboratory of Ornithology, under the leadership of Fitzpatrick, has done for the Ivory-billed Woodpecker and for the conservation of forested wetlands in the South. When other professional ornithologists, present company included, were busy with their basic research projects or more mundane conservation issues, and when virtually no professional ornithologist seemed willing to openly search for ivorybills, Fitzpatrick and Remsen stuck their necks out by organizing searches and following up on ivorybill sightings. It was a gutsy path to take. With the current frenzy over ivorybills, it is easy to forget that Fitzpatrick and Remsen started these searches when almost every authority was declaring the Ivory-billed Woodpecker extinct. Without their leadership and dedication, and without a search for the Ivory-billed Woodpecker led by distinguished and utterly credible professional ornithologists, ivorybill hunting would have remained relegated to fringe elements, discussed at conventions along with Bigfoot and UFOs. By legitimizing a continental search for ivorybills, Fitzpatrick and company finally got a few people, including my students and me, deep into the swamps.

Perhaps the greatest myth regarding Ivory-billed Woodpeckers is that *birders* would have found them if they still existed. I really find this idea amusing. I'm not ragging on birders here—birding is a hobby, and birders pursue their hobby in the most enjoyable manner. I'm ragging on the notion that birders were likely to find and document remaining populations of ivorybills. With few exceptions, birders chase and list rare birds. I should know; I'm one of them. Birders in North America spend almost zero time in wilderness, unless they are going in for nonbirding recreation. Wilderness almost never produces rarities, and by its very nature, it offers too little habitat diversity for a big list of species. Most of the birders I know don't even birdwatch in boots. They never leave roads or graveled trails. When I'm in full birder mode, working on a big day list, I spend virtually all day in the car zooming from one observation point to another. There is almost no walking involved—certainly no mud or standing water. Give the birder the garbage dump, the sewage pond, the breakwater of a marina. In Alabama, for example, virtually every dedicated birder spends his or her time moving between the winter waterfowl flocks of the Tennessee River Valley in the north and the migrants traps on barrier islands of the Gulf Coast in the spring and fall, with perhaps a few hours staking out rare hummingbirds at feeders in backyards. I don't know any birder who has ever, let alone regularly, taken a kayak into remote swamp forests looking for Ivory-billed Woodpeckers. Very few birders even drive their cars down roads near potential ivorybill swamps.

I have to conclude that virtually no remote swamp forest in the Florida Panhandle has been legitimately searched for Ivory-billed Woodpeckers since collectors last stored their shotguns 100 years ago. In his book *In Search of the Ivory-billed Woodpecker,* Jackson summarizes recent ivorybill searches, mostly by himself and John Dennis, and it hardly could be considered an exhaustive effort. Jackson never mentions so much as a day's float down the Escambia, Yellow, Conecuh, Shoal, or Choctawhatchee Rivers, all of which seem like prime spots for ivorybill populations. According to the account in his book, Jackson was unable to search the forests along the Apalachicola River because he was warned not to enter the swamp alone and could not find a guide. Timothy Spahr reported in an article for the magazine *North American Birds* that he and some colleagues conducted a 25-day search of the Apalachicola River Basin in 2003 without finding any direct evidence of ivorybills (but they did hear some suggestive sounds). Otherwise, the Florida Panhandle remains completely unsearched, and it is an immense area. I should also point out the obvious—just floating down the main channel of the Apalachicola or Escambia Rivers in a canoe or boat is not a legitimate search. Days must be spent deep in the swamp forest away from the river channel before any reasonable assessment of the presence of ivorybills can be made. It will not surprise me if we find populations of Ivory-billed Woodpeckers scattered throughout the Florida Panhandle.

Much as I like to hear accolades like "miraculous" or "astounding" used to describe our discovery of ivorybills along the Choctawhatchee River, I think that "inevitable" is probably the best descriptor. The north Gulf Coast is the part of the Southeast that has retained the largest tracts of mature swamp forest, it suffered the least pressure from collectors, it was historically the center of ivorybill abundance, and the swamps are currently subject to little duck hunting. All ivorybill searchers had to do was to question the assumptions on which earlier assessments of ivorybills were made, particularly by James Tanner, and reconsider with a more open mind where ivorybills might have persisted.

The Boynton Cutoff   6

My last visit to the swamp before the arrival of Tyler and Brian and the start of the field season was on December 2, 2005. Mark Liu and I left Auburn at 3 A.M., intending to be in the swamp at first light. Southern Alabama had a pretty substantial rain a few days before, but I thought that it was probably not enough rain to significantly raise the river.

In November I had made a brief solo trip to the Bruce Creek area and had found the swamp bone dry. With the water so low, hiking was remarkably easy. The forest floor, which in the spring had been mostly under water and had then been a slippery, muddy mess, was a smooth dirt surface. Because the area is inundated much of the year, there is almost no ground cover, and I could stroll easily through the open forest. The differences between this forest and the forests of Tuskegee National Forest, where I've done a lot of bird counts recently and which is fairly representative of the degraded forests of the northern Gulf Coast, were striking. If you tried to stroll through most areas in Tuskegee National Forest, you wouldn't get far without having to fight your way through tangles of greenbrier and black-berry. In many spots in Tuskegee, the sheer density of stems of young trees would impede you.

Along the Choctawhatchee River, in contrast, there is very little ground cover or low vegetation; mostly you find scattered patches of palmetto. The trees are large and well spaced. For about eight months of the year, the canopy is a solid mass of leaves, allowing little sunlight to reach the forest floor. And virtually the entire area is subject to months of submersion when the river rises. It is like no other forest I've known. At times it could be very beautiful, especially in the spring when wildflowers pushed up through the mud and the trees were bright and green. In the winter, however, the forest was stark and drab—brown and gray tree trunks protruding from brown and gray mud.

I was hoping that Mark and I could explore some new areas of the swamp by walking around as I had done in November. As we pulled up to Bruce Creek Landing just before sunrise, I got my first lesson in how little I understood this ecosystem. Not only had this single rain event brought the river up, it had brought the river to the highest level I had ever seen. The water in Bruce Creek (which for all purposes is the same as the Choctawhatchee River) was about 6 feet higher than it had been on my previous visit. The creek that had flowed through a steep-banked channel in November was now just a narrow, treeless path in an enormous wooded lake.

As I was to learn over the next months, the Choctowhatchee is a wild, untamed river. The Choctawhatchee River and its major tributaries have no dams. Consequently, the river fluctuates greatly in response to rainfall and is constantly either rising or falling. Except for the few oxbow lakes or very high hammocks, every section of this bottomland forest is both submerged and dry for parts of the year.

Our goal on the first day of this visit was to explore the large patch of forest across the Choctawhatchee River and northeast of the mouth of Bruce Creek. We were forced to navigate by compass because neither Mark nor I had figured out the GPS units when we set out, although Mark was fiddling with the GPS as we moved along. I had spent about 15 minutes playing with another GPS the day before, but I had been unable to master it. In most contexts I do quite well making subtle deductions. But the supposedly intuitive and obvious routines for setting and running most electronic equipment frequently baffle me. On this day, for instance, when I looked at my digital watch I had to subtract an hour. When daylight savings time had come two months before, I couldn't figure out how to change the time. "Oh well," I had sighed, "it will be correct again in April."

A typical section of bottomland forest along the Choctawhatchee River with sparse ground cover resulting from a closed canopy and frequent flooding. (Photograph by Geoffrey E. Hill.)

We spent that first day exploring both sides of the river north of Bruce Creek. We found more cavities and more feeding trees, and everywhere the forest was mature and promising. The next day we launched our kayaks before first light into Lost Lake at Tilley Landing, the infamous take-out spot on the day Tyler, Brian, and I had first located ivorybills. Our goal was to paddle south through a mile of swamp forest from Tilley Landing to Dead River Landing—the route we had failed to find at the end of our discovery day in May. This time, though, we had good maps and we knew that there was no channel between Lost Lake and Dead River Landing—it's just a couple of square miles of forest. We were hoping the whole forest would be flooded and that we could just paddle through it. We started by paddling the length of Lost Lake, which is a beautiful oxbow lake in the middle of mature swamp forest with huge cypress and oaks around it. It looked like perfect Ivory-billed Woodpecker habitat. When we reached the south end of the open lake, we plowed ahead into the flooded forest. By this time Mark had mastered the GPS and was tracking our route. He put us on a southeastern heading. For the next two and a half hours we paddled through a trackless swamp wilderness.

Mark Liu paddling down Lost Lake, a beautiful oxbow lake surrounded by mature swamp forest with huge cypresses and oaks. (Photograph by Geoffrey E. Hill.)

The oaks and cypresses here were simply enormous. The area had clearly been logged—we floated past cypress stump after cypress stump—but many huge trees had been left. I estimated that about 80% of the large cypresses had been removed, but the 20% that were left included monstrous trees. I pulled my kayak to the base of one tree, and its diameter was at least a yard greater than my nine-foot boat. It was one of the biggest trees I've ever seen. But it was not just the cypresses. All the trees were huge. There was no indication that the oaks had ever been cut, and some of these ancient giants had diameters greater than 4 feet. And this majestic forest went on and on.

The author next to a huge cypress in the trackless swamp south of Lost Lake. (Photograph by Mark Liu.)

As we floated through the forest we saw many large cavities, mostly high on the trunks of large cypresses. Most cavities were clearly very old, with thick rims of scar tissue around them, but two of the cavities we saw looked new. On one, through binoculars, you could clearly see the chisel marks where 3 inches of wood had been penetrated. Because the cavities were all high up on huge cypresses, it was difficult to accurately judge their size. We also found feeding trees with chiseled bark scattered through the forest.

After about two and a half hours of weaving our way through this forest, we emerged at a large channel about 70 yards wide with no current. We thought this must be the Dead River area, but it didn't look exactly right to me, and there were no landmarks anywhere. We wanted to mark the spot where we should reenter the forest for our return paddle to Lost Lake, but we had brought no flagging. Mark got a banana out of his lunch bag and wedged it in a fork on a branch hanging out over the water. It was surprising how conspicuous that banana was against the brown and gray winter landscape.

We paddled south down the large, calm bay and came to a place where we could see fast-moving water in a big channel ahead and a very fast-flowing small channel entering the bay to our left. After consulting our map, I decided that we were at the mouth of Dead River at the Choctawhatchee River. I concluded that the channel to our left had to be a small, nameless cutoff shown on the map. We decided to go up the cutoff to a point on the Choctawhatchee River upstream from the Dead River area. Then we could have a leisurely float back to the mouth of Dead River. We paddled up the swift channel, came out on the main river channel just as I expected, and headed south. Throughout this maneuver, Mark was studying the GPS unit with a concerned look on his face.

Mark is an excellent grad student and a great travel companion, but I have to be wary in our conversations. When I'm out on the river with Tyler and Brian, we're just three guys kicking around. Status and age don't mean much. If I say or do something stupid, they let me know about it, and if it's stupid enough, they'll make fun of me for days, the same as I would do with them. They freely disagree with me. When Tyler, Brian, and I are out in the swamp, Brian and I almost always defer to Tyler in matters of navigation. He's the best navigator of our ivorybill group by far.

Mark, on the other hand, is from Taiwan, where elders and individuals of higher rank are treated with a certain deference. For example, if I commanded Brian or Tyler, "Go get me something to drink!" they would either laugh at me or respond with something like "Get it yourself you lazy ass." I would respond in the same way (maybe a bit more politely) if my departmental chair told me to get him a cup of coffee. In Taiwan, however, it would be very reasonable for a boss to command his employee to bring him a drink, and it would simply be unheard of for a professor to pour tea for his grad student. A grad student in Taiwan would always make the tea and serve the professor. I used to have great fun with this in my lab. I'd brew a pot of tea and come over to Mark's desk, put a cup down, and pour him tea. At first it was all he could do to sit and receive it. We've now westernized Mark a lot—he has no problem with me pouring him tea. But he still has trouble contradicting me, even if I'm clearly wrong.

So even though Mark had the GPS in his hands, I have to take full responsibility for the idiotic blunders that followed. It turns out that the little channel that I called an unnamed cutoff was, in fact, the Choctawhatchee River. At this point almost the entire water volume of the river goes down a channel called the Boynton Cutoff, and the Choctawhatchee River barely looks like a creek. I could throw a rock across it left handed. One has to guess that 100 years ago the flow was reversed, but once the names (and county borders) were set, the change in flow did not lead to a change in names. I didn't recognize the miniaturized river channel, and this led to a series of idiotic assumptions and decisions. We paddled into what I thought was the Choctawhatchee River but was really the Boynton Cutoff and headed downstream, getting ourselves into a worse and worse mess with each stroke of our paddles. Mark knew there was something wrong from the start—he kept staring at the GPS and commenting that the direction we were headed didn't seem correct. But I kept confidently chipping in that there was no way we could be wrong. We'd be coming to the Dead River area any minute. He couldn't bring himself to contradict his major professor. Tyler or Brian would have been laughing at my poor sense of direction. Mark just kept looking at the GPS and frowning.

The local map in the GPS that we were using had only one landmark—the Choctawhatchee River. It showed no creeks, no cutoffs, and no landings, and we were far from any marked highways. After we had floated downstream for 15 or 20 minutes, Mark paddled over to me and said, "This shows the river that way." He pointed west.

"Well that's stupid," I said. "We're floating right in the middle of the river. That GPS map is way off." We both laughed.

If I ever again laugh at what a GPS is indicating, I invite anyone nearby to smack some sense into me. GPS units and their maps may be off by a bit, maybe tens of meters under the worst conditions, but they are never off by a mile, and given that we didn't really know where we were, it was simply inexcusable not to pay attention to what the unit was telling us. As we passed a boat ramp on the east bank of the river, it started to rain pretty hard. There was no boat ramp shown on the paper maps that we were carrying, and I didn't remember seeing one on this part of the river during our trip in May. Now you'd think by this point that alarm bells would have been thundering in my head. The GPS showed the Choctawhatchee River, where I claimed we were, a mile or more to the west, and we were passing major landmarks that weren't on the map. A little farther on there was a main channel entering the river to our left. This had to be Boynton Cutoff, I thought. "We should see Dead River on our right any minute," I proclaimed. But we didn't. We kept drifting downstream. I got more and more worried, and it started raining harder and harder. Nothing seemed right.

Finally a boat came motoring up the river. One of the nicest things about the Choctawhatchee River is that the folks who run boats up and down the river will invariably stop if they see you. They stop to help, to offer advice or directions, or just to chat. This time it was a middle-aged man in rain gear squinting into the driving rain.

"Hey, how's it going? Hell of a day to be out in boat," I started.

"It'll let up soon enough," he grunted, as one who had been in a few showers on the river.

"Are we near Dead River Landing?" I asked over the rain, which was now pounding down on us.

The man wrinkled his brow with concern. "Dead River? That's way up on the Choct. Holmes Creek is up yonder." He pointed upstream from where we had just come.

My heart sank. Way up on the Choct! Weren't we on the Choctawhatchee? Where the hell were we?

I hid my concern, thanked the man for his help, and turned away with Mark. We immediately paddled to shore, got out our soggy maps, and as we watched little rivulets of water meander across the blue and green of the map, the reality of what I had done came crashing down on me.

"Oh my God," I exclaimed to Mark. "We went down the Boynton Cutoff instead of the Choctawhatchee. We're way down here." I pointed to a spot on the map ridiculously far downstream from where we should have been.

"We have to paddle all the way up to here to get back to our car," I said, pointing to a spot on the map seven inches above where we were. I paused as the seriousness of the situation set in. "I don't know if we can do it."

We were faced with more than a mile and a half upstream paddle and the current seemed ferocious. I had joked on the way down that we were running class three rapids as we bounced over logs and debris in the fast brown current. It didn't seem like a joke now. It was just before noon. We had about five hours before dark.

The area of the Choctawhatchee River basin where Mark Liu and I got off course during a kayaking trip in December 2005.

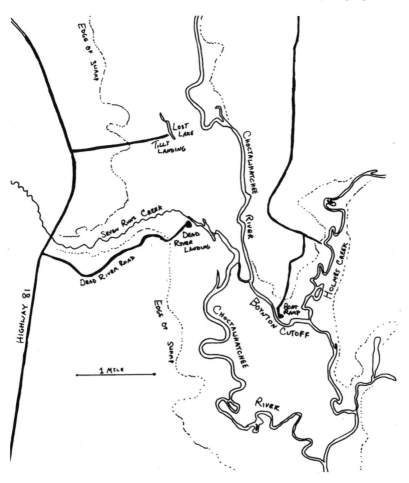

I had a sick feeling in my stomach. We had to be back to our forest passage at least an hour before dark. No way could we navigate that route through the flooded forest in the dark. That meant that we had at most 4 hours to get upstream. It was a very unpleasant prospect. For a moment I considered giving up, paddling downstream to a bridge or landing, and trying to hitchhike back to our car at Tilley Landing. But then I got my second wind. "No way am I hitchhiking back to Tilley Landing for a second time," I announced as I shook off the defeatist attitude and decided there was nothing but to put our backs into it.

It turned out that by staying very close to shore on the side of the river opposite the main current, even to the point of crashing through bushes and fallen trees, we could make pretty good time. In one place a back eddy even helped push me upstream for a few yards, but all in all it was a grueling upstream paddle. At six points we had to cross the river to escape the main current, and each crossing was a battle. About two and a half exhausting hours later, faster than I had initially dreamed that we could make it but with extremely tired arms, we were back at the Choctawhatchee River—at the little section that we had taken for a cutoff. We had a little under three hours until dark and could now relax a bit.

We backtracked using the GPS. We retrieved Mark's banana that marked our route into the forest—I've never been so glad to see a piece of fruit. We made it back to Mark's Toyota around 4:30 in rapidly dimming light and a steady drizzle. We couldn't have cut it much closer. We were driving out of the landing road enjoying the warm, dry air from the car's heater when I spotted a tree next to the road with a large patch of bright orange wood exposed. I shouted for Mark to stop and we leapt out to examine the tree.

This was the best Ivory-billed Woodpecker feeding tree I'd seen anywhere. It was a small (8-inch diameter) spruce pine dying or just dead. On the lower trunk were numerous chisel marks where firmly attached bark had been pried cleanly away from the sapwood. The top two-thirds of the tree had been twisted and bent over, apparently by a storm, and from the break point out, the entire tree had been stripped of bark. Everywhere that bark had been peeled off there were bore holes of beetle larvae. This was not loose bark just knocked off. A man could not have easily peeled that tree with a mallet and chisel. It was really impressive, and it was too far down river to have been made by the same birds that lived in the Bruce Creek area. We took a few photos, piled in the car, and headed north toward home.

Mark had generously offered to drive his car that weekend because it had racks for two kayaks, and I was still waiting for the racks for my car to be delivered. As he drove us north in a steady drizzle, Mark cranked up some bad seventies soft rock being sung in Japanese. He said this vocalist was very famous in Asia, and Mark was jamming along to the beat. With Mark absorbed in his music, I had plenty of time to mull over what had happened that day.

A recently deceased spruce pine with extensive bark scaling. (Photograph by Geoffrey E. Hill.)

How foolish I had been to get us so far down the Boynton Cutoff and so far off course. Going into the day, I thought I was gaining a bit of savvy and experience in the swamp. But I had made a series of elementary navigation blunders that nearly left us in an untenable position. As it was, I had turned a nice, relaxing ramble into a physical struggle with the river. I had allowed my expectation to shape what I saw in the way of landmarks along the river, and I had been unwilling or unable to step back and objectively examine my assumptions.

I mused how the search team in Arkansas had fallen into essentially the same trap in their woodpecker search. By the time they got to the Cache River, John Fitzpatrick and his colleagues had half a decade's experience with ivorybill searches. In *The Grail Bird*, Tim Gallagher jokes that by 2004 Fitzpatrick and a few of his confidants even had a code name for a reliable ivorybill sighting—"ground zero." They had walked away from the Pearl River in 2002 with no claims of ivorybills but with a reputation for careful, objective, and skeptical assessment of ivorybill sightings. The bird experts at Cornell knew better than anybody how easy it was to be fooled by false kents and double knocks. They were acutely aware of how a Pileated Woodpecker or even a duck or anhinga could look like an ivorybill if it was far away at a tricky angle or in poor light, and especially if the observer really wanted to see an ivorybill. They should have been the hardest individuals to bait with less than rock-solid, irrefutable evidence. And yet, they had gotten lost in the wilderness that they thought was so familiar. They had talked themselves into presenting a highly suggestive video as definitive evidence. As our own excitement about Florida ivorybills grew, I could empathize with Fitzpatrick and his team. Brian, Tyler, and I could feel the pull toward declaring what we had seen and heard as irrefutable. But it was not. I was determined not to make the same error Fitzpatrick had made in prematurely declaring conclusive proof of ivorybills. One trip down the Boynton Cutoff was enough.

Hunting Ivorybills in the Backyard   7

On the night I got back from the late May trip to the Choctawhatchee with Tyler, a few months before the start of our formal search, I was startled to get an e-mail from a birder acquaintance with "Ivory-billed Woodpecker" in the subject line. At this point Tyler, Brian, and I had not told anyone about our sighting, and as I mentioned before, I had nothing to do with ivorybill searches until our Pea River trip in May 2005. In my thirteen years at Auburn I had never received an e-mail about ivorybills, except the invitations from my friends to go to Arkansas after the Laboratory of Ornithology announcement. Why would someone suddenly be writing to me about ivorybills? The e-mail was from Charles Kennedy, president of the South Alabama Birding Association. He wrote: "I got a message today that a couple of months ago I would have sent a polite answer to and forgotten about it, but considering the news out of Arkansas I thought I would pass this one around a bit. I am sending this to you because it mentions property owned by Auburn University. If you have any interest in this you will find the contact information in the body of the message below. Probably a nut case, but who knows."

I read the message pasted below Charles's e-mail. It was a rather lengthy note from a beekeeper named Ted Kretschmann who lives in Dadeville, Alabama, only about 25 miles from Auburn. He described several recent encounters with ivorybills in Tallapoosa and Elmore counties, including a very close view of one in his yard a few years before. All of the sites that he was mentioning are well within an hour's drive of Auburn. I was very skeptical, but because my hobby was now ivorybill hunting, I felt I should follow up. These sites were north of the coastal plain and outside the historic distribution of ivorybills. But Dadeville is not far from historic ivorybill locations, and the historic distribution of ivorybills was based primarily on a handful of birds collected around the turn of the twentieth century, just before the species petered out.

I called Ted on the phone that night and spoke to him. It turns out that he was a graduate of the entomology department at Auburn University. He seemed well spoken and knowledgeable, and he was adamant that he had seen an Ivory-billed Woodpecker in his yard three years before and was now hearing them at some of his bee yard sites. I told him that I'd try to get out to his house as soon as possible.

Monday when I saw Tyler I told him about the e-mail and my conversation with Ted. Tyler was supposed to be doing point counts in Tuskegee National Forest the next day but the forecast called for rain, so we decided to go together to meet with Ted, look at his evidence, and hear his tales. We drove to Dadeville the next morning in a steady drizzle.

The best word that I can find to describe Ted Kretschmann is big. He is a big man with big voice. He is clean-shaven with a thick mop of black hair and black horn-rim glasses. He wore a black T-shirt stretched tight over his large frame, with worn jeans and hiking boots. He is probably a few years younger than I am. He spoke with only a hint of a southern accent. If I had met him at a national beekeeper's conference, I might have guessed that he was from the Midwest.

We met Ted at his little house beside a highway 10 miles north of Dadeville, where he lives by himself. I had driven by this house many times. It is on my way to Horseshoe Bend National Military Park, where I like to go birding and kayaking. Ted's house was any house and every house beside a hundred different highways in the Deep South—a modest one-story brick structure with yellow trim. It had no landscaping at all. It was just a house on a plot of cleared land. What grass was on the "lawn" seemed to be there more by accident than by intent. A dirt drive led from the highway past the house to a barn. Near the house was a wider dirt patch, which seemed like the place to park. The nearest tree was about 100 feet away, and the nearest woods were across the highway, 200 yards away.

Ted walked up as we got out of the car.

"Are you Dr. Hill?" he boomed, as he approached Tyler with his hand extended. Tyler wears a beard, while I'm clean-shaven, with a boyish face. Tyler looks more like a professor than I do.

"I'm Tyler Hicks, Dr. Hill's assistant," Tyler said, taking Ted's hand.

"I'm Geoff Hill," I said, walking around the car to shake Ted's hand. "You must be Ted."

"Yes, sir," Ted answered as he turned toward me, and then without missing a beat he was talking about ivorybills. "The bird I saw was right there on a stump of a pecan tree. I was inside the house and I saw it through the kitchen window. You can see the distance from the window to the stump. It was only 15 feet away."

I could hardly have imagined a less likely place to see an Ivory-billed Woodpecker. The blackened remnant of the stump, now burned down to the level of ground, was still visible. It was right next to his house, far from other trees. It seemed like a really crazy sighting, but Ted was adamant.

"I watched it pulling big transparent larvae out of the stump. They were *Lucanus* larvae. That's stag beetle larvae, Dr. Hill. Those beetles are as rare as the bird. I haven't seen stag beetle larvae in years," Ted explained.

"What made you think it was an ivorybill?" I asked, needing to hear how good his description would be.

"Dr. Hill, I don't think it was an ivorybill. I know it was," Ted replied in a very earnest voice. "It was right in front of me. It had lots of white on its back that formed a big white triangle. Its bill was pale cream color and its eyes were golden yellow. I was so close I could see that it didn't have any feathers around its bill or eyes."

"It didn't have feathers on its face?" Tyler repeated, making sure he understood. To this point the description had been perfect. Now it was veering toward a Turkey Vulture.

"No sir. It looked like it had a skin problem or a disease or something."

"Geeze. You must have gotten a good look at it," I admitted. It was hard to believe that someone could have identified the beetles in the mouth of a bird, seen the condition of the skin around its face, and not correctly identified it. The fact that Ted had noted an idiosyncrasy of the bird like a skin problem on the face made the sighting all the more believable. Ted wasn't just reciting what was written in field guides. He was telling us what he saw.

"Yes sir. It looked like a cartoon character. It was jerking right and left all the time. And you know I had my camera with film right in the next room. I could have gotten a great picture."

"Then why didn't you take a picture?" Tyler asked before I could get the words out.

"I never thought of it. I was just staring at the bird unable to believe what I was seeing. And then it flew off to the woods across the highway. That's old-growth forest across the road, Dr. Hill. It's about 1,500 acres and the hardwoods are 120 years old. The pines are old too but they are dying. There's an old nest hole sixty feet up in a dying white oak. It is perfectly round and around six inches across. It's an ivorybill nest, Dr. Hill."

"Wow, that's amazing," I said, "but the sighting of the bird in your yard was a while ago, right?" I asked, trying to keep all of Ted's sightings straight.

"Yes sir. That was three years ago."

"Didn't you say in your e-mail that you thought you heard one more recently?"

"Dr. Hill, that was over behind the Piedmont Substation—owned by your university. There's a huge beaver pond there that is half a mile wide and three miles long. It has old-growth timber along the valley with lots of flooded dead snags sticking out of the swamp. I have a bee yard there, and this past Sunday I heard the slow tapping of a large woodpecker. When I walked toward my bee colonies I heard this woodpecker start calling in an excited manner. I must have scared it. It sounded like someone choking a goose."

I didn't want to hurt Ted's feelings but I was thinking that his poor description of an ivorybill kent call was a lot less convincing than the bird he watched from a few feet away.

Tyler didn't pussyfoot around. "Why did you think this choking-goose sound was made by an ivorybill?"

"When I heard it, I didn't know what it was. Then I listened to Cornell's sound tracks off the Internet and the last half of the sound track and what I heard are the same. It was the hant call of a very scared Ivory-billed Woodpecker. I'm sure of it."

"You didn't see that bird?" I asked

"No sir. It was back in the swamp. I just heard it."

"What about other sightings and the nest cavity you mentioned?" I asked.

"The nest I found is closer, and I'd like to drive you over there and show you if you have time."

I looked at Tyler. He was raring to go and I was extremely curious by this point. "Sure," I responded, "let's go take a look."

We piled into Ted's big, white diesel truck and headed south. We wound our way through the back roads of Tallapoosa County, and Ted told us a bit about his business. He runs hundreds of bee yards all over the Alabama and Tallapoosa River drainages. He knows the countryside like the back of his hand, and he started telling us how little the so-called experts knew about the wildlife of the area when the topic of Black Bears came up.

"My wildlife friends tell me that the only bears in this state are in the Mobile-Tensaw Delta, and that there aren't many left even there," I told Ted.

"Dr. Hill, no offense to your friends, but I should show you some of my hives that have been tore up by bears. There are plenty of bears left in this part of the state. They are just very wary and nocturnal. They don't mind eating my honey, though," Ted explained.

"Have you told the state game folks about the bears?" I asked.

"Dr. Hill, they don't want to hear about bears from a beekeeper like me. Their minds are made up. They think its just dogs and coyotes getting into my hives. The bears are out there, though."

We arrived at the site where Ted had found the cavity and thought he might have heard ivorybills. Again, it looked nothing like I expected for an ivorybill site. We pulled into a small pasture where Ted's beehives were set up. There was maybe an acre of grass surrounded on all sides by woods. Some of the trees around the pasture were pretty big, but that's typical of pastures and hay fields throughout the state.

"This is where the nest tree is," Ted declared as he led us a few yards into the woods. The ground here fell away into a steep, wooded ravine, no doubt the reason these trees had not been cut in a long time. The oaks on the slope were big, but this woods was nothing like any forest ever described as ivorybill habitat. Ted pointed to a spot high on the trunk of a tall white oak and there indeed was a large hole cut in the living tree. Ted said it was 60 feet from the ground, and I had no reason to doubt that his estimate was correct.

During the course of the day I was extremely impressed with Ted's knowledge of the natural history of Alabama. As a naturalist I'm completely one-dimensional; my knowledge pretty much begins and ends with birds. When I first started graduate school, my obsessive focus on birds inspired one of my fellow ecology students to print a sign for my door: "If it don't have feathers, it ain't worth lookin at." That pretty much sums it up for me. Ted, in contrast, could name all the trees, shrubs, and wildflowers in the woods we were walking through. He is a professional entomologist, so he is especially knowledgeable about insects—he is probably the only person in history who has claimed to have identified both an ivorybill and the grub that it had in its mouth. His knowledge of birds wasn't tremendous, but it wasn't bad. He also understood the ecology of the area quite well.

Tyler and I didn't know what to make of this site Ted was showing us. Everything about it went against conventional wisdom regarding where the ivorybills should be. The Ivory-billed Woodpecker should have been restricted to the coastal plain in low, wet areas that held water much of the year. This site was well into the piedmont, and there could never be standing water in this steep topography. Everyone has always concurred that ivorybills need extensive tracts of forest with big trees. What we were walking in was just a finger of forest following the ravine. The trees were big, but the woods didn't seem to be nearly extensive enough for one ivorybill, let alone a pair or a population. We saw some spots on dead trees where a woodpecker had been working, including some spots where bark had been pried away, but there was nothing that couldn't be dismissed as the workings of a pileated.

After we had poked around the site for a while in a steady drizzle, Ted said he wanted to show us some other places. We piled back into his truck and drove a few miles down the highway. Ted then turned off the paved road and went down a farmer's field road to the back of a large cotton field with some standing water in the middle.

"That cotton field was a tupelo swamp eight years ago," Ted said with some remorse in his voice as we passed by the cotton field. "But they left the bigger swamp alone."

We stopped at the back of the property.

"Dr. Hill, this is only a few miles from that nest cavity," Ted stated as we stood at the edge of classic tupelo swamp. The trees were mostly of modest size, 2 feet or less in diameter, but they had broad bases and grew from standing water. At least this slough looked like ivorybill habitat.

"But the ravine woods and this swamp are separated by farm fields and young pine stands," I stated, trying to hide my skepticism but not doing a very good job of it.

"Yes, sir. I believe these ivorybills roam around among these patches of forest. At any one time there are several large patches of woods that are getting mature with lots of dead trees. When such a stand gets cut, the birds move on to the next maturing stand. I think that's how they've hung on. Moving around a wide area," Ted explained.

We walked to the edge of the tupelo swamp. This was the first place that we had seen that I thought was close to ivorybill habitat, and it wasn't great. It was a slough with mostly young and pretty dense tupelo growing in shallow water. There were only a few large trees scattered about, and the entire swamp was only 200 or 300 yards wide. Ted said it ran for a couple of miles.

After looking around the edge of the tupelo swamp for a few minutes, we piled back into Ted's truck and drove back to his house. Ted said he wanted to show us the wooded tract across from his house.

We walked into the woods directly across from his driveway on a small dirt lane that was gated at the highway. The woods were mixed hardwoods and loblolly pine for most of the way in. Some of the oaks were big, but it was a dry upland site and nothing about it seemed right for ivorybills. Compared to our site along the Choctawhatchee River in Florida, which was open with little ground cover, the floor of these woods was a dense tangle of greenbrier and blackberry. It was almost impossible to bush-whack away from the road. In most places the trees were dense, with well-developed midstory.

The road ended at an abandoned building next to a 1-acre lake. Ted followed a trail down the overgrown earthen dam that formed the lake and said as he scanned the trees below the dam, "There's a big cavity here somewhere."

In a few minutes he located what looked like a big cavity near the top of a white oak. It was hard to tell how big it was or even if it had been cut in the side of the tree or was just a spot where a branch had fallen off. Tyler and I saw some woodpecker workings on dead trees, including some places where bark had been knocked away, but we saw nothing that looked like the chiseled bark that we had been seeing along the Choctawhatchee River.

We ended the day back at Ted's house. Tyler and I promised to make some follow-up visits to look for ivorybill sign and the birds themselves. Ted said that he was going to carry a video camera with him wherever he went.

I was busy with my research for the rest of the summer, but in September I made good on my promise to return to Ted's ivorybill sites, and I spent a morning walking around the woods across from Ted's house. It was a nice morning to be looking for warblers and other migrants, and I spent a few peaceful hours in the woods. I saw no signs of ivorybills—no feeding trees, no cavities except the one Ted had pointed out. I visited with Ted briefly on my way out, and he told me that he had several more encounters with ivorybills; he had heard their kent ("hant") calls, and he had gotten fleeting glimpses of ivorybills disappearing into woodlots. He said he was sure that he would get video. I thought Ted's rate of encounters with ivorybills around Dadeville was a bit unsettling. Either the birds were here in good numbers, or Ted was getting overly enthusiastic about ivorybills and turning every glimpsed bird and odd sound into an ivorybill encounter.

Then about three weeks after my September visit, I got an e-mail from Ted. He said that he was going over an old tape that he had made in the summer at the ravine site that Tyler and I had toured. In July he had flushed a bird from the woods that he said "flew like a loon." He had zoomed his camera in on it. At the time he thought the video was inconclusive, but when he had gone back and looked at it carefully he now saw white on the trailing edge of the wings. He concluded his e-mail with: "I honestly think we have it, sir. I have reviewed the tape twice and I believe we have got an ivorybill in flight."

This note came just as we were organizing our ivorybill search in Florida. I was floored. Was it possible that Ted has gotten definitive proof of an ivorybill in the Piedmont of Alabama before we even started our spring search in Florida? Ted's note came on Friday, and I told him I could come over to see the tape on Sunday morning. He said that would be fine.

I arrived at Ted's house around 8 A.M. Ted was waiting for me with the tape in the VCR.

"This is it, Dr. Hill. I think we've really got the bird on tape," he said as he cued it up. Ted has a big screen TV, and as the camera played the tape on the TV, the tops of trees and a bright summer sky became visible on the screen. There was a tiny blurry speck in the center of the screen that the camera proceeded to zoom in and focus on. With the zoom, it was clear that the speck was a bird flying far out of the woods and up into the open sky above the field.

I stood dumbfounded. There was nothing, absolutely nothing, about the bird captured on this video that suggested a woodpecker, let alone an Ivory-billed Woodpecker. The flight did not look at all like a loon or even a duck to me. My guess, and it's only a guess because there isn't much to go on in the video, is that the bird was a Green Heron. If there was any white on the bird's wings as Ted had suggested, I couldn't see it. The bird was far away and flying away from the camera. It is hard to imagine an ivorybill flying out of a woods and flapping high into the open sky above a hay field.

"Ted, I don't think this is an ivorybill," I said sort of timidly. I didn't want to hurt his feelings, but I didn't want to lead him on.

"You don't?" Ted responded.

"No. It's flying way out into the open. The bird doesn't look at all like a woodpecker. I'm guessing it's a Green Heron."

"Oh, well. I'll have to keep at it until I get a good video," Ted responded as he turned the tape off.

I was surprised by how easily Ted gave up on the idea that this was an ivorybill. In his e-mail, on the phone, and just before playing the video, he had been ecstatic that he had a definitive ivorybill video. Now, he was ready to drop it entirely.

As I drove home from Dadeville that morning, I was left to ponder what to make of Ted's ivorybill detections. Like so many reports of ivorybills in the last half of the twentieth century, Ted's report is hard to just dismiss and ignore. I could certainly dismiss the recent fleeting glimpses and the poor video as the result of too much wishful thinking, but what about Ted's original sighting of an Ivory-billed Woodpecker in his yard right in front of him? Ted had seen details of plumage and behavior that seemed to rule out anything but an ivorybill.

Ted is a smart guy. He holds a bachelor's degree in entomology, and he is a very knowledgeable naturalist. When I was out in the woods with him he correctly identified trees, shrubs, wildflowers, insects, of course, and even many of the songbirds and woodpeckers that we saw. He strikes me as totally sincere, and I would be shocked if I ever found out that Ted had lied to Tyler and me about what he had seen. Ted truly believes that he saw an ivorybill eating stag beetle larvae in his yard.

And yet, all of that said, I remain very skeptical that Ted had an Ivory-billed Woodpecker in his yard or that he ever heard or saw an ivorybill near his bee yards. The entire context of the sightings is wrong. Ted's yard is out of the historic range of ivorybills, there is no ivorybill habitat in the vicinity, and that spot in Ted's yard is extremely exposed, next to a busy road, far from the woods or even another tree. No one has ever before reported seeing an ivorybill feeding that far outside of a forest.

So how do we reconcile my contention that Ted is truthfully describing something that he saw in his yard but that he could not have seen an Ivory-billed Woodpecker? I think that Ted saw a very strange and atypical bird in his yard. Possibly Ted saw a Pileated Woodpecker with white in the secondaries, although I have to say I'm getting tired of leucistic pileateds being used to dismiss every ivorybill sighting. A stump far from the woods would be an odd place for a pileated as well. I think it is more likely that Ted saw a Red-headed Woodpecker.

Ted reported that his bird had a skin problem that caused loss of feathering in the face. I think that skin problem may have caused the loss of all red feathers except a few on the crown of the bird. A Red-headed Woodpecker with red feathers left only on the top of its head would be a very odd looking bird indeed, and it would have a plumage pattern quite like an ivorybill. All that would be needed would be for the observer to get excited about the prospect of an ivorybill and overestimate the bird's size and fill in some details like bill and eye color. Overestimating size is easy when you see a bird much closer than normal. In discussing this sighting with Ted, he kept coming back to the white on the back as the thing that was impossible to explain. A patch of white on the back is the most striking feature of a perched Red-headed Woodpecker. Red-headed Woodpeckers prefer to feed in the open on isolated trees and stumps and are common in Ted's neighborhood. Once Ted convinced himself that he had seen an ivorybill, he caught ivorybill fever and started to hear and catch glimpses of them everywhere.

I'm sure that Ted will not like this interpretation of his sighting. Obviously, what I suggest is only an opinion. It is certainly possible that Ted had an Ivory-billed Woodpecker on a stump outside his kitchen window. If it is any consolation to Ted, I'm certain that our claim for ivorybills along the Choctawhatchee River will be similarly dismissed by skeptics as a mistake. I really hope that Ted proves me wrong some day with a clear video of an Ivory-billed Woodpecker in Tallapoosa County.

The question that I had to ask myself was how I could justify accepting Tyler and Brian's ivorybill sightings along the Choctawhatchee River in the summer of 2005 when I was willing to dismiss Ted's Dadeville sighting? For that matter, why did I tend to elevate Tyler and Brian's Florida sightings ahead of the ivorybill sighting by Tim Gallagher and Bobby Harrison in Arkansas? Ted's sighting had two big knocks against it (no pun intended)— the context was wrong and Ted had no credentials in bird identification. As unfair as it may be to Ted, putting his sighting in a different category as the Choctawhatchee or Cache River sightings is easy to justify. Ted is not a birder.

When I considered why I held Tyler and Brian's sightings from the summer of 2005 above Tim and Bobby's ivorybill sighting along the Cache River in 2004, however, I had to face my own biases and prejudices. Both

Tyler and Tim are excellent birders with extensive credentials. Tim has more years of experience birding than Tyler; Tyler, I believe, has done more guiding and run more identification workshops than Tim. Both Bobby and Brian were rather inexperienced birders with few or no birding credentials at the times of their ivorybill sightings. Tyler saw more detail on the ivorybill he sighted, but Tim got a clear look at his bird and was positive about his ID. I had to admit that there was no rational justification for elevating one of these sets of sightings above the other.

It seemed to me that the lessons to be learned from my response to both Ted's sighting in Dadeville and the sight records from the Cache River in Arkansas was that it was going to be a struggle to remain objective about the evidence we amassed. It is human nature to root for the home team. I was coming to realize that it was all too easy to point a finger of criticism at one ivorybill searcher for overstating evidence and then to turn around and be just as emphatic about one's own records. It was up to me as the head of our search team to keep us levelheaded and objective as we chased ivorybills in the Florida Panhandle.

Let the Search Begin | 8

Finally, in late November 2005, with the start of our official search effort looming, Dan and I spoke on the phone about how we would organize the search.

"My end of the search will be mostly just time spent in the swamp. I'll have the searchers move around through all of the areas near the mouth of Bruce Creek and just look and listen for ivorybills," I explained.

"Yes," Dan responded, "a random search is probably all that we can do at this point. I don't see any advantage in transects or preselected observation sites. What about cavities?"

"Oh yeah, we will watch cavities every dawn and every evening. I think we've already got ten or more good cavities to check, and I'm hoping we can find more. We'll also use video cameras to monitor cavities. Cavity watches will be a priority because I think it's our best way to get a video of a bird and to find a nest. How many listening stations are you going to set up?" I asked, curious about how Dan was going to run the sound-monitoring part of the study.

"Kyle and I will set up seven stations that can record 24 hours per day, and then Kyle will have an eighth recorder to carry with him for ad lib sound recording," Dan answered.

"Great. I hope Kyle gets lots of good double knocks and kents with his recorder."

"If he does, they should be good recordings. He's got a top-of-the-line recording setup."

"I think we should spread the listening stations around the area," I added, moving the conversation back to the listening stations. I wanted to hear Dan's ideas for station placement.

"That's what I had intended," Dan stated. "I think that we need to cast as wide a net as possible, hoping that we pick up a vocalizing ivorybill in some portion of the area. Then we can start to focus our search in areas where we get detections. We are just limited by how much distance Kyle can reasonably cover with those heavy batteries each day."

"We should definitely keep a map in camp with all of our detections plotted," I suggested. "We can also plot cavities. We probably better not try to plot feeding trees, though, because there would be so many."

"Yes, a map showing detections will help us see the areas where the birds seem to be active," Dan agreed.

"This all sounds good. It's going to be a very small search crew for most of the winter and spring. After the first week in January, it's going to be just Brian and Kyle doing the hunting, with me on site four days per month," I concluded.

"Yes, we'll have a tiny search crew. It will be almost nothing compared to the effort by Cornell in Arkansas, but I think it can work. Kyle may not be an experienced birder, but we need his technical know-how to run the stations and keep track of the huge computer files that they generate," Dan said.

"Yeah, for most of the winter and spring, Brian will be the main ivory-bill hunter. He never seems to get bored with the ivorybill hunt, and he's shown a knack for finding the birds. I hope Kyle develops into an ivorybill hunter as well," I said as we concluded the call.

In mid-December, Brian drove from Pennsylvania to Florida via Auburn, and Tyler and his dad, Leon, drove nonstop from Kansas to Florida. They were to meet in Ponce de Leon on December 17 and then set up camp down in the swamp. For a week after I said good-bye to Brian in Auburn, I had no contact with the crew in Florida. I had to assume that Brian had hooked up with Tyler and his dad and that everything was okay. On Christmas morning after my kids had finished opening their presents and I was sitting around enjoying hot coffee cake and coffee, Tyler called my cell phone.

"How're things going?" I asked, not expecting much woodpecker news yet.

"We got great sound recordings of double knocks!" Tyler blurted out, too excited to build any suspense. "Brian and I had a bird double knocking right in front of us. We turned on our cameras, and we have a bunch of double knocks recorded. Brian thinks he may have heard a second bird farther away, too."

"Did you get a look at it?" I asked, amazed that they already had what might be definitive evidence.

"No, I never saw the birds that were double knocking," Tyler answered. "But yesterday I had a bird fly right at me."

"No way! You got another look at an ivorybill?" I asked, getting more excited.

"Yeah, when I spotted it, it was flying directly at me. I was looking at it head-on. It looked like a duck—a big duck like a pintail. My first thought was 'what's a duck doing flying through the woods?' Then it banked sharply to the right as it saw me. It was only about 50 yards away. As soon as it banked, I saw that it was a big woodpecker with bright white trailing edges on its secondaries."

"What other field marks did you see?" I asked, well aware that the sightings from Arkansas had been criticized for documenting white on the wings of birds and little else.

"I saw the distinctive underwing pattern," Tyler answered matter-of-factly. "Not just a bright white trailing edge but also a white wing lining that created a black band down the center of the underwing that expanded to cover the primaries. It was a great look."

Excellent, I thought. It might be possible for a Pileated Woodpecker to have an abnormal amount of white on its wing. But no pileated with aberrant white plumage would also fly like a duck and have the white-black-white underwing pattern of an ivorybill.

"Did you see bill color or any red on the crest?" I asked.

"No," Tyler responded glumly. "You'd think at 50 yards I would have seen bill color, but I didn't. I also didn't see any red, and at that distance, red should have been conspicuous. I was mostly looking at the ventral side of the bird, though, so a red crest may have been hidden. It was flying fast, too."

Tyler was worried about losing our connection and running his cell phone out of juice, so we only talked for a few minutes. He said that for the first four days they had stayed on Bruce Creek and had detected nothing. Then they started spending more time to the south, closer to a spot marked

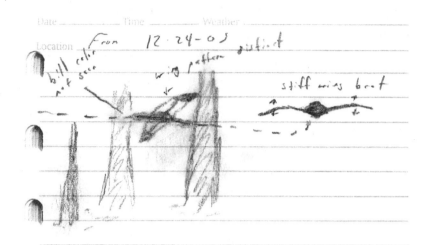

Notes made in the field by Tyler Hicks after he saw an Ivory-billed Woodpecker on Christmas Eve 2005. The underwing pattern Tyler observed is unlike the underwing pattern of any other woodpecker or any other large bird likely to occur in a southern forest.

"Story Landing" on the map, although the landing in this area had not been used for decades. This is the area in which I had heard a double knock in May and where I had located cavities during a November hike. When they shifted south of Bruce Creek, Brian, Tyler, and Leon started to detect Ivory-billed Woodpeckers. On his last day before heading back to Kansas, Leon had heard a distinct kent call, the first any of our team had heard since the previous May, and they heard a bird double knock in front of them before they could record it. The next day, Tyler had his sighting, and then Brian and Tyler recorded the double knocks that Tyler just told me about.

Brian later gave me more details about the ivorybill that double-knocked right in front of them on December 23:

"Tyler and I were walking together through Hill Swamp. All of the sudden this double knock booms right in front of us. I mean it couldn't have been 30 feet away. Tyler pointed at where the sound came from, and we both stared in that direction, but it was behind a fallen tree, and the spot was completely blocked from view. It was very frustrating. We knew that that bird was right in front of us but we couldn't see it. Then about 10 seconds later it flew off. Tyler got a brief view of a big woodpecker flashing white. It was about a 1-second glimpse as it disappeared into the forest."

Tyler said that they were camping on the high ground near the mouth of Bruce Creek, and the site seemed to be working well. As he ended the conversation, he said that they already discovered several more promising cavities and that they were going to be watching those over the next few evenings. I told him to look for Dan, Kyle, and me on January 4 after lunch.

I put down the phone and went back to the Christmas celebration in quite a festive mood. My growing fear had been that we had stumbled onto a single Ivory-billed Woodpecker or maybe a family group in early summer that had moved into the Bruce Creek area because of some food source and would move out to a distant unknown location to breed. In other words, I was afraid that what happened to the Laboratory of Ornithology team in Arkansas in 2004 would happen to us—a flurry of encounters followed by an inability to detect any birds. Now that we had detected displaying birds at the start of the 2006 breeding season, I could put that fear to rest.

On January 3, 2006, I participated in my thirteenth Auburn Christmas Bird Count as a group of one. It was a nice day to be out, and I paid particular attention to woodpeckers and especially Pileated Woodpeckers, which are common around Auburn. I had several nice looks at pileateds flying, and I studied their shape and behavior. I've been observing pileateds throughout my adult life, and I doubted I could mistake one for an Ivory-billed Woodpecker under any conditions.

I ended my count in late afternoon and drove to my house to find that Kyle and Dan had arrived safely from Ontario and were hard at work. Dan introduced me to Kyle. I'm only five-and-a-half feet tall and I felt as if I had to look straight up to say hello to Kyle, who has to be at least six-and-a-half feet tall. "It's going to make kayaking tough," I thought. "I don't know how those long legs are going to squeeze into our little boats." But Dan and I had already discussed what an advantage Kyle's height would be in setting up listening stations—getting the microphones high and the bag with the recorders far above any water. Long legs are also hard to beat when you are wading, so Kyle's height was sure to be more of an asset than a problem.

Kyle is a gentle giant. He was perfect for our project. He loves fiddling with electronic devices and computer programs, and he and Dan could fall into a techno-babble about their recorders and microphones that was as incomprehensible to me as one of Mark's phone conversations in Chinese. To run all of the sound monitors that we intended to erect and to compile all the audio files, we really needed a technophile like Kyle. When thinking about our ivorybill team in relation to the original ivorybill research team in the Singer Tract in the 1930s, I like to think of Brian as James Tanner, chasing ivorybills through the forest, and Kyle as Paul Kellogg, back in the sound truck checking the equipment.

All morning Dan and Kyle had been at the local Home Depot and an outdoor shop purchasing wooden poles and other structural components for the listening stations, as well as a canoe. Kayaks are unbeatable for moving one person through a swamp, but they are nearly useless for hauling gear. The large canoe that Dan purchased would be invaluable for hauling gear in and out of our remote camp. And we were starting to accumulate a lot of gear.

(Left) Brian Rolek and Kyle Swiston in the Choctawhatchee River bottomlands in April 2006. (Photograph by Geoffrey E. Hill.) (Right) James Tanner and Paul Kellog in the Singer Tract in 1935. (Photograph by Arthur A. Allen, ©Cornell Laboratory of Ornithology.)

Having finished their shopping, Dan and Kyle were now opening box after box of equipment that Dan had shipped to my house over the previous weeks. They spent all afternoon and evening assembling and testing the listening stations. As they opened their packages and took inventory, they realized a critical item was missing—the battery charger, which was essential for charging the batteries for the listening stations each

night. All of the other equipment was useless without it. Dan got a customer service rep on the phone and arranged to have a unit shipped overnight to my house. We couldn't wait for it, though. We were on a tight schedule and would be in Florida before it arrived. That would mean 7 hours of driving for someone to come back to Auburn to pick up the unit.

Just then Brian called my cell phone from Florida to find out the plans for the next day. Toward the end of the conversation, he happened to mention that he had set up a post office box in Ponce de Leon.

"You're a genius!" I exclaimed and passed the box number to Dan. Dan arranged to have the battery charger sent to the P.O. box in Ponce de Leon rather than my house, and a 7-hour drive was averted.

We planned to head out the next day with enough time to get everything from the landing to our remote camp before dark. That meant that we needed to be on the road by 10 A.M. By mid-morning the next day it was obvious that all the work on the listening stations could not be completed in time for a 10 A.M. departure. Dan decided that he could finish assembling the listening stations at our camp, so we loaded three kayaks (mine, Wendy's, and Mark's) on my car and the canoe on Dan's car and headed

south. Wendy was working on a grant proposal and several manuscripts and couldn't join us in our adventure. Mark was at a science meeting in Orlando. It was just as well that only three of us headed south from Auburn— we had just enough kayaks for everyone who would be in camp.

Tyler picked up Sidra Blake, his girlfriend, at the Pensacola Airport the day before and was to meet us at Bruce Creek Landing around 2 P.M. Brian intended to paddle up from camp and help us unload. Having six people rendezvous at a remote site from three different directions seemed unlikely to work, so I was relieved to see Brian, Tyler, and Sidra waiting for us when we pulled into Bruce Creek Landing. Tyler introduced us to Sidra, a quiet young woman who proved to be a fine naturalist and birder. She had just flown in from Korea, where she had taught English for the previous six months. We introduced Kyle to everyone and then set about unloading all the gear. And quite a load of stuff it was to get downstream—a

Our search crew at the onset of our formal search in January 2006. From left to right: Dan Mennill, Tyler Hicks, Geoff Hill, Kyle Swiston, Brian Rolek. (Photograph by Sidra Blake.)

big generator, 10 gallons of gasoline, 20 gallons of water, 7 listening stations, winter clothes, sleeping bags, tents, and food. Dan and Kyle piloted
the heavily laden canoe, and the rest of us each paddled a kayak loaded as
tightly as possible. Because I had brought three kayaks, there were two more
boats than paddlers. Brian and Tyler each tied a fully loaded kayak behind
his boat. We made quite a flotilla heading down Bruce Creek. The day was
gorgeous—70 degrees, sunny, windless—and we meandered our way
down the creek to camp without any problems.

Brian and Tyler had christened the main camp "Beavertown" because
they walked up on a beaver defiantly cutting saplings when they first entered. I thought that Beavertown was impressive. Already there was a covered cooking and storage area, a fire pit, and four tents.
Waders and socks hung in a neat row from poles nailed   *Kyle (left), Chet Gresham*
between trees. The food supplies and water were stacked   *(standing), and Brian (right)*
on one side of a camp table that supported a propane   *in our Bruce Creek camp that*
we called Beavertown. (Photo-
graph by Geoffrey E. Hill.)

stove where most meals were cooked. Dan, Kyle, and I quickly put up three more tents, and Brian and Tyler joked that Beavertown was now Beaver City. As we set up our tents, Chet Gresham, the final member of our search team, came walking back into camp from an afternoon search. Chet is one of Tyler's birding buddies from his teenage years in Kansas. He's now in a graduate program studying poetry in Chicago and had come in a couple of days before to help with the search. Tyler had invited two other birder friends and confidants to help with the search, but conflicts arose and they had decided not to join us.

Wearing camouflaged waders and with a week's beard stubble on his face, Chet looked like a grizzled swamp veteran. I could have easily mistaken him for one of the locals out hunting squirrels. It turned out Chet was a softy from Chicago, hardly a swamp rat. He loved birding and was greatly enjoying the chance to be part of the search for ivorybills, but the cold and wet and paddling did not seem to suit him well. When I departed a few days later, Chet was clearly ready to get out of the swamp—a week away from civilization had been plenty. But Chet has a great birder's ear, and he was an asset during the time he was part of our team.

By the time we got our tents and gear in place, the sun had set. Dan, Kyle, and I had eaten a late lunch before reaching the landing and so we skipped dinner that night. Brian started a campfire, and we sat around swapping tales and enjoying the warm evening. Tyler and Brian recounted the events of the previous two weeks. Besides the ivorybill encounters I had heard about already, they told me they had heard another bird foraging. They said it sounded "like someone hitting a tree with a baseball bat"—louder it seemed than a bird could be. We decided to start calling this loud foraging "hammering" as distinct from "double knocking." This hammering would become one of our primary methods of detecting ivorybills in our study area.

As this bird hammered, Brian turned on his video camera, pointed it toward the foraging bird, which was obscured by vegetation, and walked toward it. As Brian approached, the bird quit foraging, sat quietly for almost 20 seconds as Brian continued forward, and then quickly flew off away from Brian. Just before it flew off, Brian said it gave one loud knock—the single knock that Tanner said the birds in the Singer Tract sometimes gave when they were alarmed. Brian had gotten a quick look at the bird as it flew away from him, and he said it was a large black bird with white trailing edge to its wings. He played the video back on his camera later, hoping he had captured an image of the fleeing bird, but he didn't see it on the video. He was sure he had the audio though. This last sighting occurred across the river along a channel known as Carlisle Lakes. Tyler and Brian said there was a huge and magnificent forest on both sides of this channel, and lately they were getting most of their detections of ivorybills there.

Tyler also mentioned that one afternoon when he was coming back into camp he heard the hammering sound. He thought Brian was playing a trick on him and sauntered into camp expecting to find Brian swinging a log against a tree and laughing at him. But when he walked into camp Brian wasn't there. About 30 minutes later Brian arrived from down south near Story Landing. Whatever had made that sound was not Brian. I made everyone agree right there that we would never make jokes about ivorybill sightings. "You can play all the latrine or sex jokes you want, but we never joke around about ivorybills," I declared. "Absolutely never make fake Ivory-billed Woodpecker sounds for any reason." Everyone nodded in agreement. We all agreed that we would never play Ivory-billed Wood-pecker tapes—no one in our group had done that since we had played kent calls at the mouth of Bruce Creek one time back on our discovery day the previous May—or try to make double knocks on trees or on our boats. That way, we would never have to wonder if something that we heard or recorded was a bird or a human making a sound like a bird.

From accounts in Tim Gallagher's book and the Laboratory of Or-nithology web page, searchers in the Cache River and White River areas of Arkansas apparently regularly used playbacks of ivorybills and pounded on their boats and trees to mimic double knocks. Unfortunately, with so many searchers in the woods, including both members of the organized team and feelancers conducting their own searches, human mimicry of ivorybill sounds led to confusion and uncertainty when ivorybill-like sounds were heard by people or recorded by remote sound recording stations. The Cor-nell team was well aware of the possibility of human-created sounds, and they worked diligently to filter such sounds from their data set. We thought it was better to completely avoid such human-created ivorybill sounds.

The more news I heard from Tyler and Brian, the more excited I got. Their rate of encounters with the ivorybills was very good. The local ivory-bills were obviously shy and skittish, but they were flying around the area and making noise. We already had audio evidence of the birds, and I thought that we were bound to eventually capture one on video. With a bit of luck, I thought, we should be able to find roosts and nests.

Already, there were many cavities to watch. We had seen several intriguing cavities in our initial summer visits to the area, but now that the leaves were off the trees and Tyler and Brian began to more systematically search areas, the number of large cavities was encouraging. We informally divided cavities into three groups, A to C, ranging from those that appeared huge and newly chiseled to those that appeared old or small. The way we estimated a cavity's age was by the thickness of the woody collar around the rim. We assumed that these woody collars were a response by the tree to the tissue damage caused by the woodpeckers and that they thickened with age. Of course, cavities in dead snags didn't develop scars, but most cavities had been carved into the sides of live trees, and many of these had a thick ring of scar tissue around them. As a mater of fact, many cavities were clearly very old, with a couple of inches of scar tissues, diminishing their openings to something that a red-bellied woodpecker could barely squeeze through. We thought that some of these heavily scarred cavities could be decades old. Many other holes in trees were impressive in size, but the scarring at their lips indicated that these cavities had been carved out one or more years before. We had no idea whether ivorybills would use such older cavities, but we focused on what seemed to be the fresh cavities. Only a small fraction of all the cavities that we found looked recently chiseled with no scarring around the lip, but still we had found more than a dozen such fresh cavities. Brian and Tyler found one not very far from camp that appeared so new that the wood around the rim of the hole was starkly white. This faded to brown in the first week or so that they watched it, leading them to believe that they had found the cavity only weeks or even days after it was created.

Many of the cavities were impressive in their size and placement. Several had been cut in the sides of living oaks and other hardwoods. Cypresses were also favorite cavity trees, and some big cypresses were covered in cavities, the oldest with thick lips of scar tissue, the youngest with cleanly chiseled entrances. It was common to find trees with two or three cavities, usually stacked one above the other, but sometimes placed on alternating sides of a tree. Many of the cavities appeared to be huge—much larger it seemed than the pileated cavities we had been seeing all our lives. Some were oval, but others were roughly triangular or irregularly shaped. The largest cavities never seemed to be perfectly round. The cavities tended to be high up on big but not the biggest trees. Until Brian took up the task of measuring holes later in the spring, we didn't know the actual dimensions of any cavities.

Two large cavities in a living water tupelo. The top cavity is old as indicated by the thick scar tissue around its rim. The lower cavity is newer with less scar tissue. (Photograph by Geoffrey E. Hill.)

As Brian and Tyler told us about all the new cavities they had found, I asked about their distribution.

"Are you guys finding these cavities evenly distributed over the study site?" I asked as we sat around the fire.

"Oh, no," Tyler answered. "Some areas have lots of cavities and some have none. They are definitely clustered."

"Yeah, like Bruce Creek," Brian piped in. "We've only found one cavity along the entire creek, and it's not very impressive—it's probably just a pileated's. Also on the hammocks around here there are few cavities, even though there are lots of big trees. Bruce Creek and these hammocks have lots of feeding trees but few cavities."

By this point we were using the term "hammock" for any patch of relatively high ground that did not flood during normal high water. These patches of oak, sweetgum, and spruce pine were likely the same sorts of habitats that Tanner referred to as "first bottoms" in the Singer Tract.

"So where *are* all the cavities?" I asked, thinking about what was left besides Bruce Creek and the hammocks.

"These big cavities seem to almost always be in trees that are in standing water when the river is up—in the depressions that we are calling 'swamps,'" Tyler commented.

"Most of the cavities around here are in what we are calling Hill Swamp," Brian continued. "From right in front of camp where we pull in our boats . . ."

"Yeah, like the cypress with the 13 cavities right down there," I interjected, pointing toward the spot where we parked our boats.

". . . down to Story Landing. In that half-mile of swampy area, there have to be fifty cavities if you count both old and new. There are at least half a dozen large fresh cavities in that area."

"So that's one major cluster. Any others?" I asked.

"Yeah, definitely across the river along the lower part of the Carlisle Lakes channel. There are at least as many great cavities over there as there are in the swamp on this side of the river," Brian explained.

"Carlisle Lakes" would more appropriately have been called the Carlisle Cutoff. It is a channel of water that leaves the main stream of the Choctawhatchee far north of our study area and cuts across a bend on the east bank of the river. It runs for several miles, coming out directly across the river from the mouth of Bruce Creek. When the river is up, Carlisle Lakes can have a strong current and appears to be a substantial creek running through the forest. Brian and Tyler had been exploring the lower reaches of this area.

"And then also in the swamp behind camp that runs between Mennill Hammock and Bruce Creek," Brian added. "There are quite a few cavities in that area, too."

"And the three areas with clusters of cavities are the same areas where you've been detecting ivorybills," I noted.

"Yep." Tyler and Brian nodded.

"So we've got three cavity clusters and maybe three pairs of ivorybills," I added, always the most optimistic about the number of birds that we were dealing with.

"Has anyone ever seen so many big cavities in any forest before?" I asked the group, which collectively had birding experiences all over North America.

"Never. Not even close," was the consensus of the group. We were seeing far more big cavities than any of us had ever seen before.

"Are there that many more pileateds here?" I asked. "Can a huge pileated population possibly account for so many cavities?" My impression was that pileateds were not particularly abundant in this swamp.

"No way," Tyler chimed in. "The Pileated Woodpecker density around here seems average to me. I think the pileated density in Tuskegee Forest is actually greater, and big cavities are very scarce in that forest. A high pileated density can't explain the huge number of cavities we're finding. I guess it could be weird behavior by the pileateds that are here, but there aren't that many."

Everyone agreed that Pileated Woodpeckers were not exceptionally abundant in this swamp. Later in the winter when my friend Dave Carr visited our study site, he also commented that he was surprised that Pileated Woodpeckers weren't more abundant.

Kyle (left), Brian (middle), and Wendy (right) sitting around the campfire in Beavertown sharing observations after a long, tiring day. (Photograph by Geoffrey E. Hill.)

Sitting around the campfire sharing observations after long and tiring days would become one of my favorite activities in the swamp, and this was my first night to partake. As we continued to swap tales around the campfire, Tyler told us that he had made friends with several key local people. Probably his most important contact was a man that Tyler referred to as "William" when he first told me about him. Two days later we ran into "William" on the main river and I extended my hand.

"Hi, I'm Geoff," I said, expecting a "Hi, I'm William" in reply.

Instead, the man met my hand with a big smile and replied, "Nice ta meet cha, I'm Earl."

Tyler later explained that when they first met, the man had said, "My name's William but my friends call me Earl." Tyler had thought it presumptuous to place himself on the list of friends, so he called him William.

"I guess now we know that Earl considers us his friends," I commented.

Earl lived on the road to Bruce Creek Landing and spent every day in the swamp hunting and fishing. He was the only deer hunter we ever encountered in the swamp. He had intimate knowledge of the swamp for miles above and below Bruce Creek. He knew every channel, every low spot, every hammock. Tyler took the chance of confiding in Earl soon after he met him and told him where we were camping. It was clear that he'd find our camp soon enough anyway. Earl told Tyler that we had picked the highest ground in the Bruce Creek area, and that it was an excellent place to camp. Earl said he'd keep an eye on us, and it was nice to know that someone who really knew the swamp was watching out for us. Tyler had asked Earl what he knew about the birds in the area, but Earl answered that the birds didn't really interest him. He liked to hear them sing but he'd never learned any of their names. Too bad he never took an interest in woodpeckers—it would have been very convenient if he had been able to show us a nest.

Another of Tyler's contacts was Old Man Tuppy, a frail, weather-beaten fellow with a quick smile who seemed to be missing his lower lip. Old Man Tuppy fished at Bruce Creek Landing with his grandkids almost every weekday. It was hard to carry on a conversation with Old Man Tuppy because he was almost deaf. Tyler would ask him how the fishing was, and Old Man Tuppy would cup his hands to his ears and ask, "WHAT? WHAT'D HE SAY?" His grandkids would laugh and laugh. Mr. Tuppy warmed up to both Tyler and Brian right away and said that he would look after our cars parked at the landing. He even gave Brian fresh catfish and gar on a few occasions, so he could have fresh fish for dinner, and he was always offering him a Moon Pie, which Brian politely declined. He was quite a character, always wanting to talk and tell stories, but without a lower lip he was hard to understand.

As we sat around the fire, Tyler launched into his impression of Old Man Tuppy.

"I fishes here every day. Yes sir, I do. On Tuesdies and Thursdies I fishes with Jimmy, he's Ira's little boy, my younga granbaby. An den on Mondies and Wendsdies, I fishes with Ryan. Yes sir. Ryan is my oldest granbaby."

Tyler sucked his lower lip in as he did his Old Man Tuppy impression, and he sounded like an ancient grandfather. Most of us had never heard Old Man Tuppy, but Tyler's impression was still hilarious. Brian, who had been struggling to understand Old Man Tuppy for the past two weeks, almost fell in the fire, he was laughing so hard. It was all in good fun. Tyler had gotten very fond of Old Man Tuppy, Earl, the Chubby brothers, and all the other local people he had met in his three weeks on the site. The local people who fished and hunted the river were without exception kind and friendly and immediately wanted to help us in any way they could.

Brian and Tyler also told us that Ivory-billed Woodpeckers weren't the only supposedly extinct animal they had seen. A few days before, when they were exploring the small island at the mouth of Bruce Creek, they had seen a cougar, also called a mountain lion, jump a water channel and disappear into the forest. A small population of cougars hangs on by a thread in the Everglades area, but that's 600 miles away. Cougars were supposed to have been extirpated from the entire eastern United States, except extreme southern Florida. I was amazed that Brian and Tyler had seen a cougar just a few hundred yards from camp.

"Until we get definitive evidence of ivorybills, let's just keep that cougar sighting to ourselves," I suggested. "If we start finding too many extinct animals, we might have trouble getting people to believe us."

The next morning we split up. Dan and Kyle stayed in camp to continue working on the listening stations. I went with Brian and Chet across the river to the Carlisle Lakes area. Tyler and Brian had christened the mature bottomland forest along the Carlisle Lake channel where recent ivorybill detections had been made "Tit-ka Swamp." Tit-ka is the Seminole Indian word for Ivory-billed Woodpecker. In the two weeks they were working alone in the area, Tyler and Brian had named all the swamps and hammocks that surrounded our camp. For instance, Beavertown rested on

"Beaver Hammock." The swamp in front of camp extending south, where Tyler had seen his ivorybill, was "Hill Swamp" in honor of yours truly. South of Beaver Hammock was "Mennill Hammock" in honor of Dan, and behind that was a large, swampy area of mostly second-growth tupelo that Tyler had named "Soggy-bottom Swamp" because it was one of the muddiest places on the study site.

We got to the entrance to Carlisle Lake right around sunrise, and started paddling our boats up the channel against the strong current. A few hundred feet after we had entered the channel, a big bird flew from behind us, right over my head, about 30 to 40 feet above the water. It seemed a little bigger than an American Crow with thin neck and long bill, and it had a relatively long tail. It flew with shallow, stiff wing beats in a fast, direct manner and I didn't hear any wing noise. It was only in view for a couple of seconds, then it was gone around a bend in the channel. It was a heavily overcast morning, and in the low light of the forest against the gray sky, I saw only the silhouette of the bird—no coloration or pattern at all.

"Did you see that?" I yelled.

"Yeah," Chet replied from behind me.

"See what?" Brian called from in front me.

"A bird that looked like an ivorybill just flew over," I said, and we kept paddling against the current toward Tit-ka Swamp.

I later discussed the sighting with Chet. His look wasn't as good as mine.

"The flight was reminiscent of a cormorant," Chet commented, "but that certainly was no cormorant. It was too small. I'm not sure what it was."

The bird had gotten well past Chet before he saw it, so he had gotten only a poor view of the bird flying away. All he could say was that it was about the right size for an ivorybill with an odd, stiff-winged flight, but he hadn't made out the head or tail—just the size and flight pattern. I had gotten a better look as it flew over my head, and I was pretty sure it was an ivorybill. It looked like a big woodpecker moving up the channel with a fast, direct, stiff-winged flight. I did not put it on my life list based on that look, but I thought that it was certainly worth entering into our database as a probable ivorybill. I was hoping that I'd get a much better look at the bird before this project was over.

Brian led us to the center of the area where they had been detecting ivorybills recently, and we split up. I spent the morning tooling around the beautiful forest of Tit-ka Swamp, noting with my GPS the location of large cavities, including one particularly large and fresh cavity 40 feet up in the trunk of a living oak. I didn't hear or see anything that suggested an ivory-bill. Later that morning, Chet heard a clear double knock a short distance from where the putative ivorybill flew over us that morning, but he couldn't get his camcorder running in time to record the sound. My sighting and Chet's double knocks turned out to be the main ivorybill detections during my four days in the swamp. A lot of our time, however, was focused on set-ting up the listening stations.

From the moment we settled into camp, Dan and Kyle had been intent on getting the stations going. As we sat around the campfire the first night, Dan was sewing cloth covers for his microphones. The next day, while I was in Tit-ka swamp, Dan and Kyle spent all morning get-ting a prototype listening station set up just beyond the perimeter of camp. I would explain exactly what they were doing that took an entire morning, but I'm afraid

Dan and Kyle getting a proto-type listening station set up in January 2006. (Photograph by Geoffrey E. Hill.)

I can't. Apparently lots of technical hurdles had to be overcome. I was glad to have Dan and Kyle completely in charge of the sound monitoring. Without them on site, attempts at sound recording would have been a disaster, even if someone had given us the recorders and microphones. A person like me who can't set the time on his digital watch has no business trying to run state-of-the-art sound recording equipment.

By mid-day, when we returned from the far side of the river, Dan and Kyle announced that the first listening station was operational—set up in a nearly useless place right next to camp—but operational. They had two more stations ready to set out, and we wanted to get the next two stations up in promising locations before it got dark. Brian, Tyler, and I decided to go with Dan and Kyle to get these stations in place. We left around 2 P.M. and paddled a quarter mile up Bruce Creek to the edge of a little hammock that Brian and Tyler thought would be a good location. We strapped the 10-foot pole with the microphone at the top to a sapling that stood several meters from the nearest large tree so that the microphone would capture sound from all directions. An expensive Sennheiser microphone hung straight down in a black cloth cylinder from a foot-long bracket attached to the top of the wooden pole. The setup reminded me a bit of a street lamp. Kyle lashed the pole high so the microphone was suspended about 15 feet in the air. Dan took the cord dangling from the end of the pole and plugged it into a Marantz digital recorder with memory and battery power to run about 24 hours. Batteries and memory cards had to be swapped out every day.

The battery and recorder went into a blue dry bag that Dan had blotched with brown and green spray paint to make it less conspicuous. A dry bag is a thick plastic bag with a buckle at the top. If folded correctly before it is buckled, it is completely waterproof. It can be submerged for an extended period of time and its contents will remain dry. Not just the dry bag, but the whole setup had been painted in brown and green to make the listening stations hard to spot. More than once I walked right up on a listening station as I moved through the forest, not spotting it until I was within a few feet.

With one listening station in place, we got into our boats and paddled back past camp. We decided to put the next listening station on the far end of Mennill Hammock, in the center of recent ivorybill activity. I was just following along, taking some digital photos and paying little attention.

The microphone for a listening station at the top of a 10-foot pole. (Photograph by Geoffrey E. Hill.)

Tyler was in charge of this operation, and he directed us to a point on Mennill Hammock right behind camp where we were to park our boats.

Tyler is not just an exceptional birder; he has an uncanny sense of direction and ability to remember places he's been. Much as I was growing to love this swamp, I have to admit that it was monotonous. The topography is varied—channels, sloughs, hammocks—but these variable features are repeated monotonously. I found it very easy to get turned around and disoriented in the swamp, and even areas that I visited several times and started to think I knew could suddenly appear foreign and unfamiliar. In November when the swamp was dry, I noted how hard it would be to get lost—all you had to do was move east to hit the river or north to find Bruce Creek. With the swamp full of water, it became easy to get lost. The river was still always to the east of us, but you could not travel far on foot in any direction without hitting deep water, and likewise, you couldn't paddle a kayak far in any one direction without finding shallow water or dry ground. The swamp had become an incredible maze of water and land.

When people talk fearfully about southern swamps, they usually first mention snakes, then maybe spiders or alligators. If folks had any sense, however, they would be worried first about drowning—at least when the water was high and cold like it was this week along the Choctawhatchee—and second they would be concerned about getting lost. These are real dangers. People have an innate fear of creepy crawlies, and they elevate the risks that these little critters pose way beyond what is rational. You would have to be very unlucky to be bitten by any of these animals. Getting lost, however, is a very real danger, and, in the winter, when the days are short and the nights are long and cold, getting lost could be fatal.

Tyler seemed to experience none of my troubles in navigating the swamp. He already knew the whole area for at least a mile in every direction around camp and could find his way to most places without using his GPS. He could even find his way back to cavities or feeding trees without using his GPS, which I found to be quite a trick. Try relocating a particular tree in a vast forest. We didn't have enough GPS units the week when Chet, Sidra, Dan, Kyle, Brian, Tyler, and I were all present, and for that whole week Tyler paddled around by memory. I tended to follow Tyler like a little lost puppy. He always knew where he was and where he was going. I rarely did.

As we pulled our kayaks to shore on Mennill Hammock, Tyler showed us on a GPS that Beavertown was only 80 meters away across a heavily wooded and deeply flooded channel. It's amazing how well 80 meters can hide the sights and sounds of camp. I knew that Chet and Sidra were back in camp and that our bright yellow and blue and green tents were set up there, but we might as well have been 1000 miles away. I couldn't hear or see any sign of camp. If they had blindfolded me and marched me 100 feet in any direction, I likely could never have located camp again. Tyler knew exactly where we were, however, and set off at a brisk pace.

At our last stop the listening station had been set up right by the boats, and I just assumed that this one would be close to the boats as well. So I didn't take my jacket (which had my compass and whistle in it). It was around 3:30 P.M. and the midday heat still clung to the swamp, making my sweatshirt feel plenty warm.

Tyler set off to the south, and without thinking I picked up the dry bag with the recorder in it and tagged along with the group. As we strolled along, Tyler and Brian explained the things they had found on and near this hammock—cavities and feeding trees—and where they had heard double knocks. I hardly noticed that we were walking quite a ways from the boats. At one point we waded through a flooded patch of palmetto. At its deepest the water was over my crotch. We were all wearing waders, so it didn't matter. Brian and Tyler commented how this water hadn't been there two days before. Our January set-up weekend coincided with what turned out to be the high water point of the entire winter and spring. The river hit flood stage on the night I arrived, and water got to within about 3 feet elevation of Beavertown. We were ready to abandon camp—planning to hoist the generators and other equipment into trees if necessary—before the river crested and slowly started to drop.

Finally, after walking for about 15 minutes, we reached the spot that Tyler had picked out for the listening station. Dan and Kyle got everything set up in about 30 minutes and then Tyler told us he wanted to show us a cavity in a nearby tree. Dan still had work to do in camp, so he took a GPS and said he'd meet us back there. The rest of us followed Tyler. The tree with the cavity, it turned out, was now in a channel of deep water, and it faced away from us.

"We can easily paddle to this cavity from the other side of camp," Brian said, hinting not so subtly that it was time to give up and go home.

"No, we're right here. All we need to do is move a little farther this way and we can see it," Tyler responded as he inched forward. The water in the channel had already passed his lower chest and was moving dangerously close to the top of his waders. Tyler is not one to back away from a challenge, even a pointless one.

I was 10 feet behind Tyler, and being the shortest of the group, I was already as deep as I intended to go. Kyle was walking with Tyler, and the water was barely above his waist.

"I think Kyle's crotch is higher than the top of my head," I whispered to Brian.

"Let's look at this cavity when the water gets lower," I called out to Tyler. There were an amazing number of big cavities in our study areas surrounding Beavertown. As a matter of fact, there were so many cavities that appeared new and huge that we were overwhelmed trying to decide which cavities to watch. I had no interest in risking a dunking just to see this particular cavity; besides, Brian said we could paddle right up to it from the other side (which I did when I watched it at dawn the next day).

Finally, we found that if we moved away from the cavity along a ridge of slightly higher ground we could come around and get a good look without getting in water more than 3 feet deep. Even I could make it through 3 feet of water. The cavity was impressive: 30 feet up in the side of a living tupelo with a rim of bright white wood that looked as if it had been freshly chiseled. But with no ivorybill sticking his head out, it hardly seemed worth the time and effort we had spent to see it.

We waded back to shore. The sun was very low on the horizon, due to set in about 20 minutes. We turned and started back, but I felt disoriented. I wasn't sure where the listening station we just set was located or in which direction the boats were.

"Which way are the boats?" I asked Tyler.

"North," he answered. "Just go north and you can't go wrong on this hammock." It turned out to be extremely fortuitous that I gleaned that bit of information instead of just following along.

"OK, but you guys are the guides," I said to Tyler and Brian. "Get us home."

"Let's spread out so we can cover more area," Brian suggested. "We can walk in a line."

I meandered over to the left toward the setting sun. Kyle took a position to my right and Brian and Tyler shifted farther to the right. We started heading north. After about 10 minutes, I walked over to Kyle, who had been walking 100 feet to my right. I noticed a channel of deep water to Kyle's right.

"Where's Brian?" I asked.

"On the other side of that channel, I guess. I haven't seen him in a while."

I gave a Bobwhite whistle. That's the unofficial contact call that our group uses in the swamp. No answer. So I shouted, "Brian! Tyler!" No answer.

"Let's catch up to Brian and Tyler," I suggested, so Kyle and I tried to make our way around the channel. Very quickly things got confusing, as they are prone to do in the swamp. I knew that we had to head due east to intersect the paths of Tyler and Brian and that Tyler had said go north to get back to the boats. So we headed north and east as best we could, but our path kept being blocked by water. We had crossed just one water channel on the way in, but we were now crossing pools of water repeatedly. Some of the wet spots got too deep for our waders, and we were forced to veer more west and south to get around them.

I started to complain to Kyle.

"Why did the guys who knew where we were and where we were going take the GPS?" I grumbled. Of course, Tyler had been using the GPS to help Dan and Kyle set up the listening station, and I hadn't asked for it. I'm sure Tyler thought the GPS was unnecessary because we had only gone a kilometer from camp.

By this time we had been walking for 20 minutes or so, at least as long as we had spent walking in from our boats, and I had no idea where we were. The only plan I had was to walk north, and the setting sun was our only source for direction; we had no compass.

"As long as we keep the sun behind us and to our left we'll be heading north and we'll find our way out of this maze," I kept muttering.

And then we witnessed the saddest sunset I've experienced in a long time. Our compass sank over the horizon and disappeared. This was only Kyle's second night in the swamp, and I thought *We're going to have to spend the night out here with no jackets and no way to make a fire. This is going to be no fun at all.* Temperatures were going to drop into the low thirties that night.

We kept moving north—at least more or less north based the position of the sun before it set. We finally hit a deep-water channel that I thought had to be the edge of the hammock.

"The boats are pulled up somewhere on the edge of this hammock. I think they must be east of us still," I stated, thinking out loud. Kyle agreed. He hadn't been saying much. Kyle is never very talkative, but I think at that moment he was focused on the prospect of a night in the swamp.

It was really getting dark now, and I was wondering if we could even follow the edge of the hammock in the dark, when suddenly there was Tyler.

"Anyone want to buy a kayak?" he joked. "I found two abandoned right over there."

"You bastard!" I blurted out, half joking, half serious. "You and Brian ditched us in the swamp." I tried to act mad but mostly I wanted to hug him. We wouldn't be spending the night in the swamp.

"We didn't ditch you morons. I told you to walk north. If you want to take a tour of Mennill Hammock at night, I'm not going to stop you," he said facetiously. It turns out that Tyler had spent the previous 20 minutes hiking around blowing his whistle as he searched for us. Mennill Hammock is less than a mile long and maybe 300 yards wide. We must have been within 200–300 yards of Tyler when he was blowing his whistle, but we never heard it. It's amazing how much the swamp can attenuate sound. We were also sloshing through water most of the time, and I'm sure that didn't help. In the process of looking for us, Tyler had run into Dan hiking away from the boats. Dan had failed to find the boats with the GPS and had decided to hike back to us. It's a good thing he ran into Tyler.

So one minute we were contemplating a cold, hungry night in the swamp and the next we were paddling to a spaghetti dinner and a warm bed. Our little adventure on Mennill Hammock wasn't a complete folly. That episode drove home a few important lessons. From then on, at all times in the swamp I would have a compass, whistle, and lighter on my person. No exceptions. You never know when you will suddenly be lost and alone. I would not depend on anyone else for my basic security; I would depend on myself. I would not be caught in the swamp so totally unprepared again.

After another day of paddling around on the other side of the river with no ivorybill detections, I left on Saturday afternoon with Dan. Brian and Kyle went up to the landing with us. They needed the manual for the generator and a few other things from the cars, and Dan wanted to give them both a quick lesson in maneuvering a canoe. The three of them rode in the canoe for the trip up Bruce Creek, and then Brian and Kyle paddled the canoe back to camp. I took Wendy's kayak back home with me. It had been a Christmas present from me the year before, and she wasn't thrilled with the idea of it being left in the swamp all winter. Besides, it was bright powder blue, and all the other kayaks were green or brown. It seemed to glow in the swamp, and our operation was much less conspicuous with that kayak stowed at my house.

I left the swamp feeling very good about the project. We had the tape of the double knocks and maybe a single knock that Tyler and Brian had captured on their camcorders. Dan would take these back to Canada and convert the tape recordings into digital sound files that he could analyze.

We had three listening stations set up and had left four more stations for Tyler, Brian, and Kyle to set up in the next few days. Beavertown looked like a great location for camp. I was worried, though, that if we didn't get a video of the bird, all we would have at the end of our search would be a bunch of suggestive sightings and sound recordings.

Back at my house as Dan packed up his stuff for the drive back to Canada, I asked him at what point he thought audio evidence would constitute proof of the existence of ivorybills.

"The proof is in the pudding," he said with a mischievous smile.

"I've heard that expression my whole life. What in the hell does that mean?" I responded, a bit frustrated that he hadn't given me a clear answer.

"It means that evidence is what it is. You can brag about making the best pudding and how much better your pudding is than mine, but in the end you have to serve the pudding and let people decide for themselves," Dan continued, the smile still on his face.

"So even if we had a pair of ivorybills perched on a tree next to a listening station giving ten minutes of courtship sounds, we wouldn't have proof?" I asked, hoping he'd say it was.

"You could sell such sounds as proof. Just like Cornell sold their video as proof. Some people would embrace it. Some people would say that such recordings were just a bunch of atypical Blue Jay sounds," Dan answered, getting a bit more serious about what we needed to claim proof.

"What do you think would be the best sound evidence?" I asked, still wanting something concrete to strive for.

"I think that clear double knocks and kent calls together in the same sound clip would be hard to explain away. Double knocks can always be dismissed as atypical drums of common woodpeckers and kents can be Blue Jays, but the two sounds together would be very convincing audio evidence."

"I guess you can't beat a clear, color picture," I mumbled mostly to myself. "I'm starting to empathize more and more with the folks at Cornell and their selling of the video. It is so hard to get a picture of these elusive birds, and once you've got what you think really is an image, it would be hard to just set it aside and say 'that's not quite good enough.' The temptation to sell it as proof would be overwhelming."

"Yes, but it is a temptation we must avoid."

"Agreed. The proof is in the pudding," I added, now enjoying Dan's expression. "We'll just serve up what we've got at the end of the season."

Good Science, Bad Science, or No Science At All?   9

When I look at websites discussing Ivory-billed Woodpeckers or read popular accounts of ivorybill hunts, I'm struck by the common misunderstanding of what science is and how an ivorybill search can be made more or less scientific. Science is the process of explaining natural phenomena through deductive reasoning. Scientists use tools such as experimentation, statistics, and pattern analysis to help them make deductions, but perfectly good science, sometimes the very best science, can be done by individuals simply thinking through problems and coming to logical conclusions. Science proceeds by the formulation of hypotheses (this is where brilliant bouts of deep thought are particularly productive), followed by the careful evaluation and testing of hypotheses. Through rejection of alternative hypotheses, remaining hypotheses gain support.

A birder or ivorybill enthusiast reading this book might respond to the previous paragraph with: "Thank you Professor Hill, but what does this little lecture about science have to do with ivorybills?" Exactly. When we hunt for ivorybills, whether we are distinguished professors or blue-collar laborers, whether we keep careful notes or never record anything, whether we are skilled birders or awful stringers, we are not doing science. We are not trying to explain processes of nature. We are searching for a bird. We are birding.

I think many birders are scientist wannabes and view science with a certain reverence, implicitly equating science with an activity that is particularly worthwhile and relevant. In truth, many birder-related activities that are not science, like construction of breeding bird atlases and winter bird counts, are very worthwhile and have an important and immediate application, and many scientific investigations are trivial and will probably never matter much to anyone. Equating the importance of an activity as to whether or not it is scientific has led to some sadly misdirected efforts. I have met several bird banders who have devoted their lives to putting little silver rings on as many birds as possible. They do this with a single-minded devotion sometimes driven by the mistaken idea that they are doing good science and that their activity will lead to conceptual breakthroughs in our understanding of birds. They often gather specific, essentially random, information about the birds that they are banding (weight, fat class, wing length, etc.), which they dutifully record in notebooks and maybe even enter into a computer database. These banders are adamant, even arrogant, in their assertion that they are doing science and advancing our understanding of migration or population biology or whatever. But how is bird banding science? You might just as well walk out the front door of your home with a notebook and start randomly recording characteristics of your neighborhood—the diameter of trees you encounter, the average height of grass on the lawns, the percent of blue sky that is visible each hour of each day of the year. If you kept careful records of these observations, it is conceivable that someday in the future some scientist could use your data to test a hypothesis, just as an ornithologist might someday use banding records to test a hypothesis. But the act of randomly collecting data, however interesting, is not science.

Back to ivorybills. Where is the science in a search for a rare bird? Is it possible to make an ivorybill search scientific? No. But what if we use really fancy and complicated equipment? Still no. If a population of ivorybills is discovered, then conservation or population biologists can use science to understand the habitat use and other aspects of the bird. For instance, there is speculation among ornithologists that ivorybills use upland pine stands as regular feeding areas, particularly if the pine stand has been burned recently. This idea could be formulated as a hypothesis, and then from this hypothesis predictions could be generated, observations could be made to test the predictions, and possibly even experiments could be performed. If our hypothesis is that ivorybills feed on the trunks of burned loblolly pine, then we could burn a stand of loblolly adjacent to ivorybill woodpecker territories and predict that we will detect the chiseled-bark feeding sign typical of ivorybills on the trunks of loblolly. But such scientific investigation would begin only after a successful ivorybill search.

The search itself is not a scientific endeavor. No hypothesis about how nature works is being tested. What about: "I hypothesize that the Congaree Swamp has a population of ivorybills." Isn't that a scientific hypothesis? No. If we let that sort of statement stand as a scientific hypothesis, then everything becomes science: "I hypothesize that I can hold my breath for two minutes." The word "science" would lose its meaning.

I am not denigrating the value of descriptive natural history or the importance of searches for rare animals such as ivorybills. Just because these activities are not science does not mean they are not worthwhile or that they could not be vital preambles to important scientific study. Indeed, a period of observation is a very important phase in the formulation of hypotheses leading to scientific investigation. Obviously, no scientific study of ivorybills could commence before the bird is proven to exist.

If ivorybill hunting is not science, then why are scientists conducting the searches? They aren't. Scientists might be coordinating searches, functioning as administrators, but they are not the primary searchers. Most ivorybill searchers in the Cache and White River areas in Arkansas, and certainly all of the good searchers, are birders, not necessarily scientists. Brian and Tyler at our Choctawhatchee River site, in my opinion the two greatest ivorybill hunters on the planet, have ambitions to become scientists, but at present they are birders, not scientists. At the start of the search, Kyle was a much better scientist than Brian, but he was a much less proficient ivorybill hunter. Being a scientist does not make a person more or less qualified to search for ivorybills. It is irrelevant. The scientists who are searching are also typically birders, and they are in birder mode, not scientist mode, when they are in the swamps.

What scientists can bring to an ivorybill search is a logical, deductive approach to assessment of evidence that ivorybill hunters accumulate. Some lay observers confuse logical deduction with science. Logical deduction is a foundation of modern science, and scientists are trained to think in a logical, deductive manner, but such deduction is also used in many human endeavors outside of science. Plumbers and auto mechanics, for instance, use a deductive approach to troubleshoot problems. I think that the closest analogy to the use of deductive reasoning in an ivorybill search is the use of such reasoning in a forensic investigation. A skilled detective investigating a crime scene must assess evidence and draw conclusions using logical deduction, just as the supervisor of an ivorybill search might assess evidence (video, audio, cavities, feeding sign) and draw logical conclusions regarding the assertion that Ivory-billed Woodpeckers are in the area. An ivorybill hunt works best with a mix of scientists and birders. Birders seek out and accurately identify birds by sight and sound. Scientists organize and interpret the evidence that birders bring back.

Here is a great myth that I almost hate to dispel because I benefit directly from it: professional ornithologists are the best-trained field ornithologists and are the best qualified to assess sight records and video evidence for the existence of ivorybills. This statement is absolutely untrue. A hundred years ago it would have been true. Up through the early years of the twentieth century, virtually all the individuals with the best skills at identifying birds, dead or alive, were professional ornithologists associated with museums and universities. Amateur birdwatchers as such did not really exist. By the end of the twentieth century, however, the ornithological world had transformed completely. Professional "ornithologists" had become systematists, behaviorists, physiologists, or ecologists, often with very limited skills in bird identification. At a modern university, no scientist advances in his or her academic career because he or she is skilled at identifying birds in the field. Professional zoologists, whether they study birds or other organisms, are successful because they have skills at deductive reasoning, effective writing, statistical analysis, modeling, and so on. If a professional ornithologist trained in the late twentieth century has skills at identifying birds in the field (and a handful of professional ornithologists are top birders), it is invariably because he or she taught himself or herself to identify birds or was taught by amateurs, typically before starting grad school.

Consider my training, for example. I am a proficient birder. I know all of the birds in eastern North America by sight and sound, but I gained none of these skills during my twelve years in college (four undergrad, three master's, five doctorate). The vast majority of what I know about identifying wild birds in North America I taught myself between the ages of fifteen and twenty-two. As a teenager, I pursued birding completely in a vacuum. I had no one to guide me or to answer my questions. I simply had a weird, inexplicable, insatiable interest in everything having to do with birds. No one else in my extended family has the slightest interest in birds or any topic related to natural history. My children could not care less about birds or wildlife, despite a bit of prodding early on by me. As a bird fanatic, I am truly an oddball.

I started "birding" down the barrel of a gun. Not a shotgun—I grew up in Kentucky in the suburbs of Cincinnati, Ohio, and we didn't go blasting shotguns in our suburban backyards. I started with a BB gun. (Actually I started with a slingshot, but try hitting a bird with a slingshot. I only ever killed one, a Golden-crowned Kinglet.) My parents' little suburban yard was surrounded by houses with big picture windows, so I didn't shoot much there. But my father and grandfather were partners in a golf driving-range business. This driving range was bordered on two sides by a major tributary of the Licking River, Banklick Creek, and had acres of forest and fields behind it that were ideal for a young bird enthusiast to explore. My best friends, Mark "Tilly" Tillman, Chuck "Chuckwagon" Hitter, Dave "Quarr" Carr, and I, "Little Hill," all got BB guns when we were twelve and thirteen, and we roamed around the driving range shooting stuff, especially birds. We liked and respected wildlife and birds, so we developed a bird-shooting ethic to be sure we didn't overharvest any species. In retrospect, as twelve year olds we instinctively had a better hunting ethic than the museum men and collectors pursuing ivorybills in the early twentieth century. We had no knowledge of real game laws, and we had no idea that songbirds had legal protection. We made up our own rules.

The author at age 11 fishing along Banklick Creek with Mark "Tilly" Tilman (left photo) and Dave "Quarr" Carr (right photo). (Photographs by Edward Hill, Jr.)

We put special value on "big" birds, so we prohibited the shooting of anything larger than a robin. Robins were abundant, and we decided that they could be killed without restriction. The same was true for starlings and "chippies." A chippy was any small, brownish bird. Today I know that chippies included everything from Tufted Titmice to Song Sparrows to Eastern Phoebes, but to a twelve year old these were just unidentifiable little brown birds. At this point I owned one general natural history volume that had pictures of about twenty North American birds, and my parents had a set of Collier's Encyclopedias published in the 1950s with a few bird pictures. These were my only references for bird information and identification. I vividly remember the first unknown bird that I ever identified to species, a Rufous-sided Towhee that Tilly brought down with a spectacular shot as the male sang from the top of a tree at least 80 feet away. We used iron sites on our BB guns, so hitting a little bird at 80 feet took considerable skill. When we found the towhee under the tree bleeding and gasping for breath, it was clear to us that this was not an ordinary chippy. Its bold pattern of black, orange, and white plumage looked so distinct that we thought we must be able to find a name for this bird in a book. Luckily, among the twenty or so birds pictured in my natural history book was a painting of a "ground robin," a colloquial name for a Rufous-sided Towhee. It was a perfect match for the towhee, and for the first time I realized that I could use books to identify the birds that we were shooting.

We blasted our way through the next year, trying to identify birds when we could. Then, two events in the summer of 1974 changed my life. First, my mother, who was a high school English teacher, took me to a faculty picnic at a local park. One of the faculty members was Eugene Blakenbaker, a biology teacher who had an interest in natural history, including birds. Mr. Blakenbaker announced that he was leading a birdwatching walk, and he had some extra binoculars, which he lent to me, while he carried a spotting scope. My little sister Nancy and I trudged along behind Mr. Blakenbaker as he led a short bird walk. The bird walk was merely 30 or 40 minutes in the middle of the afternoon, and we found only a few species. But to me it was an epiphany. A bird sound rang out, and Mr. Blakenbaker told us matter of factly that the bird was a Carolina Wren. I was astounded. He hadn't even seen it, and yet he knew what it was. A little farther along, a dark bird perched in a tree over the road. Tilly and I would have pronounced it a chippy and tried to drop it with our guns. But Mr. Blakenbaker looked at it with his binoculars and told us it was a Brown-headed Cowbird. A cowbird? I had never heard of such a thing. How could he know? He brought the bird into focus in his spotting scope and let me look. And there it was, a bird with a glossy black body and brown head. Through the scope, I could see it in great detail. And then Mr. Blakenbaker got out a book that also astounded me. This book did not have a few birds listed. It had *every* bird pictured and named. The implications were huge. With that book I could sort through the chippies we shot and give them specific names.

I returned from the bird walk and told my mom in an excited voice that Mr. Blakenbaker had a book that showed all the birds and that you could see them so well with binoculars that you could call them by name. My mom saw the spark in my eyes and she knew what she had to do. Despite a tight family budget, within a week she presented me with a special gift— a pair of 10 × 50 Jason Empire binoculars and *The Golden Guide to Birds of North America* by Chan Robbins. I remember my excitement when I first took my binoculars and bird book into the field.

For a few weeks I continued both to hunt and birdwatch but then, one afternoon, Tilly and I were walking with our guns near a small pond behind the golf range. A Green Heron flew up from the pond and into a tree. Our little hunting club had strict rules against shooting birds like herons, and we had no intention of shooting this bird. It was probably 200 feet away and hidden in the leaves of the tree as we walked toward it. For no particular reason, Tilly swung his gun up at his hip and shot toward it. It was just a stupid kid thing to do. The bird was so far away that with careful aim we would have had almost no chance of hitting it, and Tilly didn't intend to shoot the bird. But Tilly's hip shot nailed the heron, and to our horror the bird tumbled out of the tree, bleeding and gasping.

Tilly was beside himself.

"Oh my God, oh my God. I didn't mean to hurt it. You think it will be okay? I didn't mean to hurt it," he stammered.

But it wasn't okay. It was mortally wounded, and we watched it die on the ground in front of us.

I felt bad for Tilly. I felt bad for the bird. I felt bad for myself. That would be one less really neat bird for me to look at with my new binoculars.

From that point on, I vowed to never kill another bird (a vow I sadly broke a hundred times over as a professional ornithologist with a collecting permit in my hand). Nevertheless, our senseless slaughter ended, and I started to chase birds exclusively with my binoculars.

Over the years, I got better and better at bird identification, but I never met another serious birder until I went to college. I only had a vague concept that there were other people in the world interested in birds—I had never met such a person except Mr. Blakenbaker. For five years, I birded completely in a vacuum, matching birds to pictures in bird books. I learned bird sounds by chasing down the birds and studying how they sounded.

When I got to graduate school, my birding skills put me light years ahead of other students who showed up excited about studying avian ecology but who were incapable of telling a wren from a sparrow. This is why I think it is so important to distinguish between the skills of a birder and the skills of a professional ornithologist. They are not the same thing. It was birders, led first by Roger Tory Peterson, who took the concept of identifying birds in the field and created a manual for how to do it—Peterson's famous field guides. Peterson and his fellow birders in the mid-twentieth century refined and legitimized sight records as a reliable means to document the presence of a bird species in a location. Before amateurs proved that it was possible to definitively identify birds afar by sight, all verifications were made by way of a shotgun. Through the last decades of the twentieth century, a growing expertise in field identification of birds was developed and refined by skilled amateurs. The most proficient field ornithologists in North America and Europe are now mostly amateurs. I don't want to denigrate my fellow ornithologists who are top field biologists and are among the most skilled people at bird identification. I just want to make it clear that there are many scientists who work on birds who have, at best, modest skills at identifying birds.

The lay public is used to turning to professionals in all technical matters. If cell phone communication suddenly becomes erratic, we turn for an explanation to professional physicists whose expertise is solar flares and extraterrestrial radiation. We don't ask amateur sky watchers. If we fear an avian flu pandemic, we turn for answers to professional virologists and epidemiologists. We don't ask midwives or even our local pediatrician. Likewise, when a claim was made that a supposedly extinct bird had been observed, the public turned for answers to professional ornithologists. Only this time, they would have been wise to also listen to the opinions of top amateur birders, like David Sibley and Kenn Kaufman, in considering the ivorybill evidence coming out of Arkansas. Scientists are skilled at assessing how logically compelling a case for ivorybills might be, but they are likely to do a poor job actually assessing the details of a rare bird report. Top birders are keen at assessing sounds, descriptions, pictures, and videos but generally not very good at assessing the validity of data or the strength of a logical argument. Both skilled birders and professional ornithologists have a role to play in assessing a claim of ivorybills.

The Big Woods Conservation Partnership, led by John Fitzpatrick and the Laboratory of Ornithology, made their announcement about the existence of an Ivory-billed Woodpecker along the Cache River in Arkansas in a short article published in the journal *Science*. *Science* is one of the two top science journals in the world—the other being the British journal *Nature*. It was *Science* and *Nature* that announced such scientific milestones as the structure of DNA, life on Mars, and three-and-a-half-million-year-old human footprints in Africa. As I explained, a search for ivorybills is not a scientific investigation, but the existence of ivorybills is of general interest to many scientists. *Science* was an appropriate place for the announcement of the rediscovery of the Ivory-billed Woodpecker. It is very important, however, for nonscientists to understand just what was involved in getting

The Science *cover announcing the discovery of an Ivory-billed Woodpecker in Arkansas. (American Association for the Advancement of Science.)*

---

a paper into *Science* and the implications of such a report in a scientific journal.

A paper in *Science* or *Nature* is a crowning achievement to which all professional scientists aspire. In an age when biology is focused on understanding genetic codes and biochemical pathways, it is especially hard to get reports of ecology, evolution, or conservation biology published in these top journals. Many leading ornithologists have productive research careers without ever getting a paper into *Science* or *Nature*. The scientists whom I know only submit their very best, most exciting work to *Science* and *Nature*—maybe two or three submissions during their entire careers. Despite receiving only the highest quality papers, *Science* and *Nature* reject most of the manuscripts that they receive. The large majority of submissions are declined within 48 hours without being peer reviewed. Among the select few reports that are sent out for peer review, only a small fraction are accepted for publication.

The paper announcing the rediscovery of ivorybills in Arkansas had Fitzpatrick as first author, followed by a long list of coauthors from The Big Woods Conversation Partnership. Before he officially submitted the ivorybill manuscript, Fitzpatrick contacted the editor-in-chief of *Science*, Donald Kennedy, asking if the journal would be interested in publishing a definitive record of an ivorybill. (Fitzpatrick actually called it an "iconic" bird species, but Kennedy understood what he meant.) In an editorial, Kennedy said he responded to Fitzpatrick's e-mail "in a New York second." I am extremely impressed, and more than a little envious, of the clout Fitzpatrick demonstrated in his dealing with *Science*. To send a personal e-mail to Donald Kennedy, editor-in-chief of *Science*, and get an immediate personal response seems akin to sending a note to President Bush and getting an immediate personal response. This analogy is obviously an exaggeration, but I can tell you that no scientist that I know has personal correspondence with the editor-in-chief of *Science* or *Nature*. My colleagues and I feel honored if we get any sort of a personal note from subeditors at one at these journals. When I submitted to *Nature* the paper summarizing our evidence for ivorybills along the Choctawhatchee River, all of my correspondence was with the assistant to the subeditor in charge of Brief Communications—a person three or four tiers below the editor-in-chief. I didn't even rate a personal note from a subeditor. Fitzpatrick's access to Kennedy means that in April 2005 he was power a broker in the world of science. As the director of a major scientific institute at an Ivy League school, Fitzpatrick's political clout should not be surprising. His dealings with Kennedy at *Science* were not typical for an ornithologist, but "not typical" in no way implies inappropriate.

The rest is history. Kennedy sent the paper out for peer review, decided it was of sufficient merit and interest to deserve space in *Science*, and published the announcement. The fact that the Arkansas ivorybill announcement appeared in such a prestigious journal made many amateur bird enthusiasts think the proclamation was bullet proof. Didn't peer review by such an elite publication make the truth of the article virtually certain? In reality, Fitzpatrick's paper was in no way a consensus opinion of the ornithological community. As with all technical reports, it simply expressed the opinions of the authors.

Publication in *Science* was also not a declaration by the journal that there was proof of an ivorybill. Fitzpatrick or some folks at the Laboratory of Ornithology may have used "proof" or "conclusive" in subsequent interviews, but these words do not appear in the *Science* article. Scientists don't deal in proof and certainty. They deal in ideas and deductions. For instance, we will never have proof that Ivory-billed and Imperial Woodpeckers shared a recent common ancestor. We might get stronger and stronger evidence to support such a claim, eventually reaching a point that scientists are virtually certain it is true. But a statement of the evolutionary relationships of two bird species will always be based on logic and deduction, not on direct observation, and it will never be a truth. Ivorybill hunting is not science; we *can* directly observe these birds and have definitive proof of their existence by way of a clear and detailed photograph or video clip. But as a science journal, *Science* is not in the business of declaring proof. By publishing Fitzpatrick's article, *Science* declared "Here is an interesting report that we feel has sufficient merit to warrant space in our journal." Every paper ever published in *Science*, *Nature*, or any scientific journal is subject to second-guessing and criticism. Criticism of published accounts does not mean that journals are making poor decisions about what they publish. Assessment, reassessment, and criticism are the means by which the scientific community keeps investigations moving in the right direction.

An interesting footnote to the Cornell announcement is that it was not the first time a supposedly extinct bird in the United States was pronounced "rediscovered" in *Science*. The November 1, 1918, edition of *Science* included the headline, "Alleged rediscovery of the Passenger Pigeon." In the article, John Clark, the director of the New York State Museum, conveyed a letter from a man in Amsterdam, New York, who said he observed a flock of Passenger Pigeons flying from a buckwheat field, noting in particular their whistling wings. In 1918, the Passenger Pigeon had been gone from the wild for about two decades, and observation of a flock of these birds was not unlike spotting an Ivory-billed Woodpecker in the twenty-first century. In a critique published in the next issue of *Science*, T. S. Roberts pointed out that Mourning Doves, not Passenger Pigeons, produce a whistling sound with their wings and that Passenger Pigeons more commonly fed on mast in forests than on grain in open fields. This claim of Passenger Pigeons faded into history along with the bird.

I've heard some outcry that *Science* should retract Fitzpatrick's article in light of subsequent criticisms. Such suggestions are nonsense. Journals retract articles if they discover fraud or an error in a calculation. Journals do not retract papers because some readers disagree with the deductions that are made. Failure to find ivorybills along the Cache River in 2005 and 2006 does not mean that the bird in the 2004 video is not an Ivory-billed Woodpecker, nor does it mean that sight records from 2004 were mistakes. Critics are free to publish critiques and reanalyses, and if the new arguments are stronger than the original claim, the original claim will lose support.

I have also heard cries that peer review failed in the case of the Arkansas ivorybill announcement. Again, I don't think this is true. Let's consider how peer review works. The editor of a journal is challenged with making sure that new and original ideas, which might challenge dogma in a field, get fair consideration and are not censored while at the same time serving as a gate-keeper so that ill-conceived, flawed, and sloppy work is not published. To attempt to achieve this difficult balance, all reputable journals from *The Wilson Bulletin* to *Science* rely on peer review. Generally, reviews are solicited from between two and five experts, and exactly how reviewers are chosen is completely up to the editor. Kennedy was challenged with finding reviewers with knowledge of Ivory-billed Woodpeckers, bird identification, sight records, and video analysis. The reviewers were probably asked to give an opinion on the paper within 48 hours. If I had been a reviewer of Fitzpatrick's paper, I likely would have recommended publication even if I had doubts about the identity of the bird in the video. Unless I could be certain that the authors were mistaken—and even after a year of intense scrutiny by a large number of devoted skeptics no one has convinced me with certainty that Fitzpatrick and his colleagues were wrong— I would feel obligated to allow the authors to express their opinions. If submitted manuscripts were rejected every time a reviewer's opinion differed from the author's opinion, nothing but consensus would ever be published. I think that *Science* was correct in publishing the paper. *Science* should also be commended for subsequently publishing dissenting opinions so that the lay public did not mistakenly interpret the rediscovery paper as above reproach.

It is important to take the *Science* article for what it is—a statement by Fitzpatrick and his co-authors that they have a video and sight records of what they are convinced is an Ivory-billed Woodpecker. It is not a consensus opinion of the ornithological community. It is not a statement of irrefutable truth.

Placing the evidence for the existence of an ivorybill in the scientific literature was an appropriate course of action. But as an ivorybill enthusiast, I also want to know what top bird identification experts think of the video and sight records. From the day of the announcement, the Internet was abuzz with opinions that the bird in the video was not an ivorybill. Here and there, we also got quotes in news stories indicating that some top birders did not agree with Fitzpatrick's interpretation. Finally, David Sibley and a few colleagues published a formal challenge to Fitzpatrick's interpretation of the Luneau video in *Science*. What I wanted to see was a thorough review of a complete dossier of evidence from the Cache River by expert birders.

Evaluation of a rare bird report is done best by rare bird (aka records) committees, which are usually run by top amateur birders. The best rare bird committees are composed of top identification experts who typically have decades of experience assessing the validity of rare bird sightings. State committees are highly variable in how carefully they assess records—committees in the top birding states such as California and Texas tend to be excellent, but committees in some states are not. Many of these committees have carefully crafted guidelines regarding what does and does not constitute definitive evidence of the existence of a bird. Having a rare bird sighting rejected by a rare bird committee does not mean that the bird was not there; it just means that the evidence presented failed to rise to the level of definitive as specified by the clearly stated guidelines of the committee and the expert judgment of the committee members.

The Arkansas Rare Bird Committee, composed mostly of ornithologists who are also birders, voted four to one to accept the ivorybill sightings and video from the Cache River. To my knowledge, a summary of the basis for this decision has not been published. Decisions by rare bird committees are virtually never published (typically almost no one cares), so there is nothing unusual about an assessment of the Cache River sightings and video not being publicized. Given the level of national interest and the amount of money and number of people involved in the Arkansas search, I think that it would be worthwhile for the Arkansas Rare Bird Committee to publish its written assessment.

What about the organization responsible for assessing records of rare birds of national significance—the American Birding Association (ABA)? What is the opinion of the ABA now that the Cache River ivorybill sightings and video have been made public? The ABA has not stated an opinion yet. The status of the Ivory-billed Woodpecker in the ABA checklist has always been classified as code N (6). "N" denotes a "Native breeding species." Code 6 is an oddity. It is defined as "cannot be found. The species is probably or actually extinct or extirpated from the ABA Checklist Area." Certainly extinct species such as Labrador Duck and Passenger Pigeon are listed in the ABA Checklist with the designation E (6), where E denotes "extinct." The ABA has not changed the status of the Ivory-billed Woodpecker in light of the reports from Arkansas, but as of October 2006, no one has submitted a dossier of evidence for Ivory-billed Woodpeckers in Arkansas to the ABA Rare Birds Committee. As an aside, I would note that the Ivory-billed Woodpecker was never listed as extinct by the U.S. Fish and Wildlife Service only because at a meeting of the Ivory-billed Woodpecker Advisory Committee in 1986, Jerry Jackson cast a cautionary vote in opposition to James Tanner and Lester Short, who were ready to pronounce the bird extinct.

My take-home message from this ramble about science and birding is that I see no indication that anything inappropriate took place with publication of evidence that there was an ivorybill along the Cache River in Arkansas in 2004. Fitzpatrick and crew may be wrong about ivorybills in Arkansas, but they published their evidence in an appropriate manner. I think that the Laboratory of Ornithology should send a dossier to the ABA records committee, but there is no rule that says it must. Lots of people sit on rare bird records for a lot longer than a couple of years and no one cries "conspiracy" or "cover-up."

Likewise, for the past six decades, there has been no conspiracy among ornithologists to suppress ivorybill sightings. Many people seem to have the impression that scientists, including ornithologists, work by consensus. I hear things like "Jackson broke rank" when he wrote his critique of the evidence for ivorybills from Arkansas, as if professional ornithologists held secret meetings and agreed to let the Arkansas sightings stand unchallenged. As a professional scientist, I can assure you that scientists are anything but supportive and protective of each other's work. In science, when it comes to defense of ideas, it is "every man for himself." Ideas last only as long as they can be defended. As a scientist, it is one's duty to challenge ideas that are seen as flawed.

There is a perception among bloggers and ivorybill enthusiasts that professional ornithologists have engaged in a conspiracy to suppress the rediscovery of the ivorybill. The dust jacket of the most widely read book on the Ivory-billed Woodpecker, Tim Gallagher's *The Grail Bird*, begins with the premise of a cover-up: "Since the early twentieth century, scientists have been trying to prove that the ivory-bill is extinct. For decades every sighting has been met with ridicule and scorn." Gallagher would have us believe that he and a small band of rebels fight to reveal a truth that scientists have been working to suppress. All he needs is Dana Scully by his side and a back-alley encounter with the smoking man, and this fantasy would be complete.

The particular episode that is at the root of much of these conspiracy theories occurred at the 1971 American Ornithologists' Union (AOU) meeting. George Lowery, a well-respected professor of ornithology at Louisiana State University, came to the meeting with two photos of a bird clinging to the side of a tree. The bird in the pictures was small and a bit blurry, but it appeared to be a male Ivory-billed Woodpecker. The only inconsistency in the picture was that the white patch appeared to extend a bit too far up the bird's back. Lowery showed the photographs to some fellow ornithologists in the hallway at the meeting and got some skeptical responses to the pictures. I have no first-hand knowledge of this exchange (at the time I was an 11-year-old shooting birds with Quarr and Tilly), but Gallagher describes what happened in detail in *The Grail Bird*. As a professional ornithologist who has since attended many AOU meetings, I think that I can add some perspective. Gallagher writes on page 100 that "Lowery's pictures were met with immediate withering skepticism by most of the other ornithologists." This sentence clearly gives the impression that ornithologists as a group ganged up on poor Dr. Lowery, who was only trying to pursue the truth of his photos. But on the next page of Gallagher's book there is an excerpt from a letter by George Lowery to James Tanner—a first-hand account of what happened—in which Lowery writes that "several people expressed the opinion that the bird in the photographs is a mounted specimen." "Several people" is not "most of the other ornithologists." "Several people" in this context sounds to me like a few opinionated and, I would imagine, loud and aggressive ornithologists. A few likely candidates immediately spring to mind. Gallagher then concedes, "Part of the problem was that he [Lowery] refused to divulge the man's identity or even where the picture was taken." I guess this would be a problem. If you were an ornithologist at a professional meeting and one of your colleagues showed you some snapshots of a purported ivorybill, but the bird looked a little funny in the pictures and your colleague refused to tell you where the pictures were taken or by whom, would you be skeptical? A little caution and skepticism are not bad when you're dealing with a bird that, in 1971, had not been documented in about thirty years.

If Lowery had seriously wanted these pictures to be considered as evidence for the existence of an Ivory-billed Woodpecker in Louisiana in 1971, he should not have used conversations in the hallway as his means for disseminating the evidence. He should have carefully documented all aspects of the encounter with the birds (the observer actually reported seeing two ivorybills when the photo was taken), including not just the photos but also the other sight records by the photographer. Much to his credit, Tim Gallagher tracked down the man who took the photos, Fielding Lewis, and interviewed him for *The Grail Bird*. In his interviews with Gallagher, Mr. Lewis stated that he had seen ivorybills several times in his years hunting ducks and kicking around the swamp before he photographed them. Had Lowery properly compiled these records and submitted the record to a bird journal or to a rare bird committee, the history of ivorybill searches might have been quite different. Failure to have these photos properly considered as evidence of ivorybills is completely the fault of Lowery; there was no conspiracy to suppress information, just some skeptical scientists who weren't impressed by what they saw in the hallway at a meeting.

It's interesting how the level of scrutiny to which evidence is subjected seems proportional to one's level of personal involvement in the evidence. When Fitzpatrick gave a plenary address at the 2005 AOU meeting on the Ivory-billed Woodpecker, he put Fielding Lewis's snapshot of the ivorybill up on the screen and briefly discussed the 1971 incident. Fitzpatrick intimated that he was one of the skeptics who had rejected the photo at the 1971 AOU meeting as inconclusive, and he said that he hadn't accepted the photo as definitive evidence because the white on the back seemed too extensive. Almost in the same breath, Fitzpatrick showed the audience the Luneau video with an image of a bird that was strikingly poor in relation to Fielding Lewis' photo. In contrast to the unmistakable plumage pattern of an ivorybill in Lewis' photo, there were virtually no clear field marks on the blurry, indistinct bird in the Luneau video. In the few frames in which the video camera captured the image of what is purported to be an ivorybill perched on a tupelo trunk, white appears to extend too far up the back for a typical ivorybill. Nevertheless, Fitzpatrick concluded that the bird in the video is an Ivory-billed Woodpecker. It's hard to understand how white plumage that appears too extensive could have led to the outright rejection of Lewis's photo but have been considered acceptable on the Luneau video.

Fitzpatrick's inconsistency in scrutinizing the pattern of white on purported ivorybills underscores just how hard it is to stay objective when dealing with evidence of ivorybills. It is extremely difficult to gather evidence for the existence of these shy and wary birds, and there is a nearly irresistible temptation to accept less than definitive evidence as conclusive. As a result, there is the director of the Laboratory of Ornithology, one minute playing the skeptic and pointing at a bit of extra white on a clear photo of an ivorybill as he dismisses the record and minutes later showing his own blurry video of a bird with what appears to be too much white and glossing right over it. It seems to me that the discrepancy was made completely subconsciously, but it speaks volumes about how expectation can guide observation. During our search, as we gathered evidence, I found myself and other members of my team accepting things that we saw, heard, and recorded as definitive while at the same time dismissing evidence of the same quality coming from other sources. Clearly, it was going to be a challenge for us to stay objective as we were drawn deeper and deeper into our hunt for ivorybills.

Tangible Evidence   10

By June 2005, Tyler, Brian, and I knew that there was at least one ivory-bill near the mouth of Bruce Creek in the Choctawhatchee River bottomlands. Brian and Tyler had seen it, and I had heard it. But Alexander Sprunt, Jr. had also *known* that there were ivorybills along the Santee River in South Carolina in the late 1930s. John Dennis had *known* that there were ivorybills in the forests along the Chipola River on the Florida Panhandle and in the Big Thicket area of Texas in the 1970s. John Forsythe apparently had *known* that there were ivorybills in swamps in southern Georgia and at undisclosed locations in Florida in the 1950s. The lore of the ivorybill is replete with sightings that cannot be substantiated and that eventually become part of the legend of the bird. Our challenge was not to let this happen with our ivorybill encounter.

Just a couple of years earlier, Tim Gallagher and Bobby Harrison had been in the same situation we found ourselves in. They were sure they had seen an Ivory-billed Woodpecker along the Cache River in Arkansas, but as Gallagher says in his book *The Grail Bird*, they needed tangible evidence: "We need a picture or a video. Believe me, even if the director of the Lab of Ornithology had been in the canoe with us, he would say the same thing. We've got nothing."

They may have had nothing in the form of tangible evidence for the existence of ivorybills, but their connections to the Laboratory of Ornithology at Cornell meant that they had a pipeline to enormous resources for immediate follow-up field work by a large crew of experienced ornithologists.

Quoting again from Gallagher's book, once John Fitzpatrick was convinced that Gallagher and Harrison had indeed sighted an ivorybill, he said that "studying the ivory-bills in eastern Arkansas would now be the number-one research and conservation priority at the lab." Fitzpatrick has been a dedicated conservation biologist throughout his career, and the thought of finding and perhaps saving a remnant population of ivorybills was certainly the foundation of his enthusiasm. No doubt he also realized that documenting a population of ivorybills would be a tremendous achievement for the Laboratory or Ornithology and that such a discovery would justifiably earn the lab prestige and admiration. All they needed was indisputable tangible evidence of one ivorybill, and it would be the ornithological discovery of the century (at least for North American ornithologists). From the onset their goal was a clear photo or video of one bird.

When, two months after the original sighting, ivorybill searcher David Luneau captured a tiny and fleeting image (soon known to everyone as the "Luneau video") of a black-and-white bird flying into the woods along the Cache River during continuous videotaping from a camera mounted in his boat, it seemed miraculous. I'm sure Fitzpatrick and the rest of the search crew thought that good fortune was shining on their effort. In retrospect, though, the Luneau video may loom as one of the most unfortunate things to ever happen to the Laboratory of Ornithology or to the conservation of the Ivory-billed Woodpecker. The Luneau video was the turning point in the whole Arkansas ivorybill affair, and it was not a turn for the better. Without the Luneau video, the Laboratory of Ornithology would likely have remained as conservative as they had been during the Pearl River search in 2002. Instead, in my opinion, they were drawn into claiming too much based on too little.

The Luneau video was just good enough to show a black and white bird fleeing into the forest; it was just poor enough to show no field marks clearly. Like a fluffy white cloud in a summer sky, the image in the video could be pretty much whatever you chose to see. And Fitzpatrick, the Cornell researchers, and thousands of birders around the country really wanted the bird in the video to be an ivorybill. The Luneau video proved a temptress too alluring, and so the suggestive video image became definitive evidence for a living ivorybill. "Conclusive proof" Fitzpatrick called it in the video press release on the Cornell web page. The news was proclaimed in the pages of the most prestigious journal for scientific discovery—*Science*. Next to Fitzpatrick at the press conference was Gale Norton, the secretary of the interior, and Scott Simon, regional head of the Nature Conservancy, the most powerful nongovernmental conservation group in North America. The press conference seemed to be a statement to the world that not just Cornell University but the Department of the Interior and the Nature Conservancy accepted the video as definitive proof of an Ivory-billed Woodpecker in eastern Arkansas.

I don't mean to be overly critical of Fitzpatrick and the Laboratory of Ornithology regarding their announcement of the rediscovery of the Ivory-billed Woodpecker. By and large, I think Fitzpatrick and his team deserve high praise for their dedication to the search for remaining populations of Ivory-billed Woodpeckers and for their tireless work to preserve the southern swamp forests that ivorybills inhabit. The sightings by Bobby Harrison and Tim Gallagher and other members of the search crew and the video captured by David Luneau are unquestionably very intriguing. No one can dispute that there is substantial evidence that a male Ivory-billed Woodpecker was in the Cache River area in the winter and spring of 2004, and I understand why the groups

The press conference in Washington, D.C., in April 2005 announcing the rediscovery of the Ivory-billed Woodpecker in Arkansas with John Fitzpatrick at the podium. (Courtesy of Cornell University.)

that sponsored the search were excited by this evidence. Whether intentional or not, however, the press conference and subsequent statements by prominent search team members led the birding public to believe that ivorybills were known with absolute certainty to be in Arkansas. In my opinion, there was insufficient evidence to make such a definitive claim.

In January 2006, Jerry Jackson published a harsh critique of the Luneau video and the Arkansas ivorybill rediscovery in the world's leading bird journal, *The Auk*. As I mentioned earlier, Jackson was not just some Johnny-come-lately looking to get publicity at the expense of Cornell. He had spent his long career studying woodpeckers in North America and was widely acknowledged as the greatest authority on both Ivory-billed and Pileated Woodpeckers. In Jackson's opinion, the video purporting to show an Ivory-billed Woodpecker was not just inconclusive, it was a video of a Pileated Woodpecker. A month later, David Sibley, generally recognized as the greatest authority on bird identification in North America after the death of Roger Tory Peterson, was first author on a rebuttal paper in *Science* that reanalyzed the Luneau video and concluded that the bird in the video was likely a Pileated Woodpecker. To be fair, I don't think Jackson's or Sibley's arguments for the image being a Pileated Woodpecker were any more convincing than Cornell's arguments for the image being an Ivory-billed Woodpecker—we all see what we want to see in the Luneau video. That's the problem. It is a very intriguing video image, but I don't think it is definitive evidence for the existence of an ivorybill.

Furthermore, Jackson mused that sound recordings from the White River area purported to be double knocks and kent calls made by a resident ivorybill could be lots of things, only one of which was a living ivorybill. It is entirely conceivable, Jackson argued, that the sounds were made by people playing tapes of ivorybills or otherwise imitating ivorybill sounds. They might also be atypical drums of common woodpecker species and kent-like calls of Blue Jays. There was, in Jackson's opinion, only weak evidence for the existence of one ivorybill in the Cache River area. I found Jackson's criticism of the sounds recorded by Cornell unfair. The sound experts at the Laboratory of Ornithology painstakingly compared their sound recordings to Blue Jays, the old Singer Tract recordings, and a huge range of potential sources other than ivorybills before they tentatively concluded that the most likely source of the sounds was Ivory-billed Woodpeckers. If Fitzpatrick and the Laboratory of Ornithology are to be criticized for their portrayal of the Luneau video, they certainly deserve credit for being very careful and conservative in how they interpreted their audio evidence.

Even if a bit overstated in places, however, Jackson's critique was much needed. The Arkansas evidence fell short of definitive, and before Jackson's critique many birders had gotten the idea there was unanimous support among ornithologists for Cornell's claim of ivorybills. I think it was a bit shocking for some birders to get their first glimpse at how academics freely argue opinions.

Watching events unfold in Arkansas gave my research group a roadmap of problems to avoid in Florida. When we were ready to publicly state that we had found ivorybills in Florida, in addition to submitting our observations and analyses to a journal for publication, we would seek outside opinions from experts like Jerry Jackson to ascertain whether our evidence was definitive. We would ask a rare bird committee to review our evidence. If a reasonable argument could be mounted against our evidence, we would present the evidence that we had as suggestive, not definitive. We were determined not to oversell weak evidence and back ourselves into an untenable position. We were, of course, determined to gather evidence that would be universally accepted as definitive. As the Cornell group already knew only too well, obtaining definitive evidence of a living Ivory-billed Woodpecker was much easier to dream about than to accomplish.

My research team had four likely sources of tangible evidence of ivorybills: cavities, feeding marks on trees, audio recordings, and video footage. Cavities constituted by far the most conspicuous evidence that there was a big bird with a very strong bill living in the flooded forests along the Choctawhatchee River. The most obvious and numerous large cavities were on the trunks of huge, old cypress with diameters of 3 feet or more. Many of these old cypress had no marks, but if such a tree had one cavity, it almost always had many cavities. One large cypress in our study area had 21 large cavities, and many had more than a dozen. We weren't sure what to make of these huge cypress covered in cavities. The entrances seemed to be access holes to the hollow centers of these large, ancient trees. When we finally had a chance to measure some of these cypress cavities, some were more than 5 inches in diameter, and they were invariably roundish, although the rims of most were not perfect circles. Many tended to look rather square.

There are no animals in the forests of the southeastern United States that can chop large cavities in live trees except woodpeckers. The only

A large cypress near Bruce Creek with numerous old cavities. (Photograph by Geoffrey E. Hill.)

woodpecker in North America that is close to the Ivory-billed Woodpecker in size is the Pileated Woodpecker. Dan searched the literature and found surprisingly few published dimensions for pileated cavities. James Tanner cited an old reference stating that the entrance holes of pileated cavities were "3 to 3.5 inches" in diameter, and these dimensions match pretty well the few other published dimensions of pileated cavities. In contrast, the entrances to the three Ivory-billed Woodpecker nest cavities that were measured in the Singer Tract averaged 5.4 inches tall by 4.3 inches wide. Ivorybills excavate substantially larger cavities than Pileated Woodpeckers, and big cavities should be a good source of tangible evidence for the presence of ivorybills.

In his *Birds of North America* account of the Pileated Woodpecker, which is based primarily on studies of Pileated Woodpeckers in Oregon, Jackson writes that pileateds frequently use large hollow trees as roost trees, and when they do, they cut multiple entrances so they can escape if a pred-

ator gets inside the hollow tree. We thought that maybe the cavity-ridden cypresses were the work of pileateds, but I had never before seen anything like these numerous large cavities cut into living cypresses, and I had visited many cypress stands around the South over the years. The old literature suggests that cypress was a favored nest site of ivorybills in Florida. The nest that

Cavities in dead sweetgums along the Choctawhatchee River that are large and irregular in shape, reminiscent of the cavities of ivorybills in the Singer Tract in Louisiana. (Photograph by Geoffrey E. Hill.)

Arthur Allen and his wife watched in Florida in the 1920s was in a live cypress, and Allen wrote that there were other holes in nearby cypress with rims of scar tissue, apparently from past years. We actually hoped that the cavities in huge cypress on our study site weren't made and used by ivorybills because there were so many of them that it would make finding an active roost hole daunting.

Most other cavities on our study site were in hardwoods, including water tupelo, several species of oak, sweetgum, and red maple, as well as in solid, mid-aged cypress. These cavities seemed to be big and varied in shape from close to round to quite oval. Many had irregular shapes reminiscent of cavity entrances of known ivorybill nest holes in the Singer Tract. They often occurred as pairs or triplets on the same tree, stacked one on top of the other. Trees with cavities also tended to cluster in specific water channels. There were large sections of the swamp with few or no large cavities and other sections with six or eight cavity-trees within a few hundred feet. Large cavities most often seemed to be cut in trees that stood in water for much of the year.

Cavity trees had a mean diameter at the height of the cavity of about 2 feet, ranging from 10 inches to 5 feet. The majority of large cavities were cut in the trunks of living trees, but some big cavities were cut in dead trees. We found a handful of cavities excavated in old, rotten snags on high dry ground in the hammocks, and we thought that these were likely the work of pileateds. The cavities in dead trees that interested us were carved near the tops of recently dead trees in standing water, most often in sweetgum that still had bark attached. These trees were particularly intriguing because some bore bark chisels indicative of ivorybills lower on their trunks. All of these types of cavities varied greatly in how high they were placed on trees. Most cavities were placed between 25 and 35 feet above the ground, but some were as low as 18 feet or as high as 90 feet.

I had never before seen cavities like these—multiple cypress cavities, huge holes in solid living tupelos and oaks, huge holes in recently dead trees. I've been kicking around woods and birding in the Southeast my whole life. For the last seven years I've spent a few dozen weekends a year paddling rivers around Alabama. I even paddled the Little Pee Dee River, Waccamaw River, and Congaree Swamp in South Carolina specifically looking for cavities. Rarely have I seen a single large cavity carved in the side of a live hardwood in any of these southern river basins. By the end of our first winter along the Choctawhatchee River, we had found about 200 cavities that appeared to have entrances at least the size typical of pileated cavities in our 2-square-mile study area. If we were just dealing with pileated cavities, then pileateds of the Choctawhatchee River basin chopped more and larger cavities than pileateds anywhere else in the Southeast.

Cavities in live hardwoods often occurred in stacks of two or three, as in these cavities above my head in a live water tupelo near Bruce Creek. (Photograph by Wendy R. Hood.)

I told Brian that one of his priorities was to measure cavities. He first tried "placing" an object of known size near cavity entrances for perspective. It turned out to be very tough to get an object attached to a string 40 or 50 or 60 feet up in a tree close to a cavity. Brian tried throwing, casting, and shooting a weight up near cavities, but it was a total failure. We needed a different approach.

In February I searched the Internet for a long retractable pole and found the Wonder Pole, a white plastic pole that is 6 feet long when retracted but

extends to 40 feet. I brought the Wonder Pole down for Brian in early March and he immediately began to measure cavities. Brian drew a ruler on the end of the Wonder Pole in black marker. He would then extend the pole, lash it to a cavity tree so the ruler was next to a cavity and get back as far as possible and photograph ruler and cavity. It was then a simple matter to use computer software to measure the height, width, and area of cavity entrances from the digital images. We also recorded the diameter of the tree at cavity height. Brian got as far back from the cavity trees as possible when he photographed them, trying to make the angle between the camera and the cavity as shallow as possible. Still, we knew that photographing cavities at an angle from below would slightly distort their size. The distortion should be to make the vertical dimension smaller, so we considered our estimates of cavity size to be conservative.

The 40-foot plastic pole that we used to extend a rule adjacent to cavity entrances so we could estimate their size. (Photograph by Brian W. Rolek.)

Measuring cavities was exhausting work. It involved hauling the heavy pole through the swamp to the cavities, hoisting it into place, and maneuvering to get the best angle for a photo. The effort was paying off, though. By the end of our spring search, Brian had measured 131 cavity entrances of the approximately 200 that we found, and his results were encouraging.

Twenty of the cavities that Brian measured were more than 5 inches high and an astonishing 67 were more than 4 inches wide. Clearly, many of the cavities we found were consistent with ivorybills and were bigger than those pileateds normally excavate. Our data confirmed our impression that these were exceptional cavities and gave us tangible evidence for ivorybills. It also made sense that many cavity entrances were between the size that we expected for typical pileated cavities and typical ivorybill cavities. Most of the cavities that Brian measured were clearly cut in years past because they had thick lips of scar tissue around their entrances. Such lips would be expected to reduce the size of cavity entrances. We imagined that many of these cavity openings started out larger than 5 inches in vertical dimension but shrank in size. The cavity data were far from conclusive, but it seemed like quite a coincidence that just where Tyler, Brian, and I spotted big woodpeckers with white trailing edges to their wings, we also found numerous giant cavities bigger than had been reported for pileateds.

In addition to the cavities, we found numerous feeding trees on the study site. These were less dramatic than huge cavities, but they were very intriguing. As I mentioned earlier, these trees had not been "peeled." Peel-

Bark scaling on a large dead hardwood revealing the bore hole of an insect. (Photograph by Geoffrey E. Hill.)

ing makes me think of taking the skin off of a banana, where the outer layer easily comes away from the inner layer. There were plenty of rotting, dead trees on our study site with big sheets of loose bark. Old decayed trees were frequently worked on by woodpeckers, but such trees weren't what attracted our attention. The feeding trees that caught our eyes from the first day in the swamp were scaled or, as I like to say, chiseled; a powerful bill had pried very strong, tightly adhering bark off those trees.

Sometimes we would see where a single rectangle of bark about one inch square had been struck from the side of the tree, and usually in the center of such a chisel there was a beetle bore hole. On some trees, though, we'd observe that a large area of tightly adhering bark had been pried away one small section at a time. When an entire tree was thus stripped of bark, it did look like the tree had been peeled, but "completely chiseled" might be a better decription. On such trees each small section of bark had to be pried loose with substantial force. The pile of bark below a chiseled tree would all be in small pieces. Tightly adhering bark does not come off in big sheets.

Many species of trees showed bark chiseling, but the majority of trees that we observed with such scaled bark were sweetgum and spruce pine, the only species of pine that grows in this swamp. We have never watched an Ivory-billed Woodpecker feeding in the Choctawhatchee River basin, so all my interpretations of feeding sign and my assertion that some feeding sign was made by ivorybills were based on descriptions of ivorybill feeding sign reported by Tanner as well as my own comparisons to woodpecker feeding sign that I have seen in other southern forests where ivorybills do not occur.

The ability of ivorybills to scale tightly adhering bark is apparently made possible by their bill morphology. Unlike Pileated Woodpeckers, which have bills that are essentially round in profile right to the tip, the bill of an ivorybill is somewhat flattened at the tip. This flattened tip allows an ivorybill to use its bill as a human would use a wood chisel. When driven by powerful neck muscles, the relatively flat bill of an ivorybill can very efficiently penetrate between even the most tightly adhering bark and the sapwood beneath. Woodpecker species with rounder bills are very good at chopping holes into trees, even through hard bark and wood, and they can certainly separate loose bark from a dead tree. But round bills are poor tools for separating

A close-up of the bills of two Ivory-billed Woodpecker specimens showing the flattened tips. (©Steven Holt/VIREO.)

tightly adhering bark from recently dead trees. We thought that the marks left on trees by the broad, chisel-like bills of ivorybills were distinct from the narrower bill marks of pileateds. I like to think of the difference between the bill marks of pileateds and ivorybills as the difference between the marks that would be left by an ice pick versus a wood chisel. I should note that in his book *In Search of the Ivory-billed Woodpecker*, Jerry Jackson alludes to the relatively flat tips of the bills of Ivory-billed Woodpeckers based on his inspection of numerous museum specimens, but no biologist has yet conducted a careful study of morphology to confirm the assertions of bill morphology that I've made.

We found evidence that pileateds sometimes worked on tightly adhering bark. I watched one pileated fly to the base of a live overcup oak and in two places strike the bark away. But when I examined the spot where the pileated had fed, it didn't look like a bark chisel. It was a gouge. Where the pileated had struck a hole in the tree, it had penetrated the bark into the sapwood. I think that the round bill of a pileated makes such a gouge unavoidable. What we were seeing on chiseled trees, in contrast, were places where bark had been separated from sapwood cleanly, with no gouging of the sapwood. In one of the most dramatic feeding trees that I found, the bird that had been feeding caught a bit of the sapwood under the bark and curled a paper-thin section of wood. I couldn't imagine how the round bill of a pileated woodpecker could curl wood in this manner.

The most conspicuous and abundant feeding sign of pileated and other woodpeckers were gouges in the rotting wood of long-dead trees. Pileated Woodpeckers commonly made huge excavations into such soft, rotting wood, often moving impressive amounts of material. I guessed that ivorybills also fed this way, and Arthur Allen describes ivorybills in Florida chopping to the center of rotten snags to get at beetle larvae. As a matter of fact, I found several small dead trees that had literally been chopped in half by a foraging bird, and I thought that these felled trees were probably the work of ivorybills. I thought that many gouges into the soft wood of rotten trees could be the work of either ivorybills or pileated and likely sometimes both. But the chisels on hard bark seemed distinctive, and I thought that such scaling could be used as tangible evidence of ivorybills if we could put numbers to the distinctiveness.

The question was how to quantify what seemed to me to be unique chiseled-bark feeding marks. I pondered the idea of measuring the hardness of the wood exposed by the chisels, thinking that more rotten trees would have softer wood. But wood hardness was not what was unique about these trees. The wood of some dead trees gets very hard as it dries, even as the bark falls away in sheets. I decided that what was unique about the putative ivorybill feeding trees was how tightly bark adhered to the edge of a scaled section. I started to ponder how to measure bark adhesion.

I couldn't find any reference for measuring bark adhesion on trees, so I invented my own device by modifying a digital scale. This scale was sold for measuring the weight of fish up to 30 kilograms (66 pounds). It came with a wire hook made to slip into the gill of a fish, and I replaced the hook with an L-shaped bracket made of steel (a picture-hanging bracket). To measure bark adhesion, I would tap the lower part of the flat metal L under the bark adjacent to a feeding chisel and see how many pounds of force were required to lift the bark 1 centimeter. My daughter Savannah and I tried out this bark adhesion measurer in a neighborhood park. It seemed to do a good job of dif-

A tree with a diameter of about 6 inches that was chopped in two by a woodpecker. (Photograph by Geoffrey E. Hill.)

My 12-year-old daughter Savannah helping me try out my bark adhesion measurer. (Photograph by Geoffrey E. Hill.)

ferentiating between loose bark and tight bark, so I started measuring bark adhesion in our study area along the Choctawhatchee.

When I was measuring bark for our ivorybill study, I treated each tree as an independent sampling point. I didn't just selectively measure the most interesting feeding trees. Within a patch of forest, I measured every tree with bark that showed signs of woodpecker activity, regardless of whether it was very old and rotten or still alive, and I included trees as small as 2 inches in diameter and as large as 4 feet in diameter. My goal was to make such measurements across our study area where we had putative ivorybill feeds and then in mature swamp forest where we knew that ivorybills did not occur—the forests along Sougahatchee, Uphapee, and Choctafaula creeks near Auburn. (We were confident that ivorybills did not occur in the bottomland forests along these creeks because Tyler and I had done exhaustive bird surveys in these woodlands for years.) I predicted that if these bark chisels were indeed the work of Ivory-billed Woodpeckers, then the adhesion of bark on some of the feeding trees in our study site would be significantly greater than the adhesion of bark on the feeding trees in areas where Pileated, Red-bellied, and Hairy Woodpeckers were common but ivorybills were absent.

The results of my bark adhesion measurements were encouraging. One interesting thing was that there was a lot more bark scaling in the forests around Auburn than I remembered or expected. I had never paid careful attention to woodpecker feeding trees before, but when I conducted systematic searches, I found lots of places where woodpeckers fed by removing bark from trees. Some of the feeding trees in the Auburn region looked a lot like our putative ivorybill feeding trees along the Choctawhatchee River. There is no way to really capture the difference in photographs. When I started looking carefully and pulling bark, however, the differences were clear. Woodpeckers around Auburn do not scale tightly adhering bark off of trees. They wait until the bark is loose and then flake it off. I found no exceptions among the hundreds of trees that I examined and measured. Occasionally Pileated Woodpeckers will feed on trees that still have pretty tightly adhering bark, but invariably in such cases they gouge a hole into the rotting wood of the tree. Away from the Choctawhatchee River, I have yet to see a tree on which tightly adhering bark was scaled.

In bottomland forests around Auburn, the greatest adhesion of bark on a woodpecker feeding tree was about 9 kilograms (19 pounds) of resistance, and this in a huge, dead loblolly pine on which the bark was loose but so thick that it took a substantial pull to raise it away from the tree. Only a few other trees in the forests around Auburn had resistance greater than 5 kilograms. Almost all the woodpecker feeding trees around Auburn had bark resistances of 2 kilograms or less.

In contrast, in our study area along the Choctawhatchee River, I measured many feeding trees with bark that gave greater than 10 kg of resistance, including some small trees. Some of the "best" feeding trees (meaning trees that I would have pointed out to visitors as likely ivorybill feeding trees) had bark at the edge of woodpecker feedings with more than 28 kilograms (that's more than 60 pounds) of resistance. It was a workout pulling against this tightly adhering bark. The chiseled feeding trees that looked different and interesting to us were quantifiably different from woodpecker feeding trees in areas known to lack ivorybills.

Of course, my greatest hope for definitive proof of an ivorybill was the same as that of every other ivorybill hunter—a clear photo or video. Our strategy for obtaining a video of an ivorybill was simple—spot a bird and film it. Through December, January, and February, we realized how hard this would be. Ivorybills along the Choctawhatchee River are shy and skittish birds that flee from people. We heard ivorybills regularly, and by March

2006, Brian, Tyler, and I had sighted ivorybills on twelve occasions where we didn't have time to turn on and raise a video camera or where the video camera had not captured an image of the fleeing bird. I had to admit that Brian and I were pretty terrible videographers. I would only give us an even chance of getting a decent video of an ivorybill if it perched for 30 seconds within 100 feet. Even with our poor abilities with video cameras, given our high rate of encounters, it seemed to me just a matter of time before some-one got video of a bird perched on a distant tree or flying in front of a boat that had a mounted camera.

In retrospect, we should have outfitted all the boats with video camera mounts so that everyone could film continuously as they paddled around the area. We knew that this was how David Luneau had gotten his famous video along the Cache River. After I had an ivorybill fly in front of my boat in January, I finally ordered a scope mount made for a car window think-ing that I could use it to attach a video camera to the rim of my kayak. When I got the mount, however, I realized that the slot for attaching to a car window was too narrow to fit over the rim of my kayak. "No problem," I mumbled to myself as I got out my pocket knife. With a few quick slices I narrowed the rim of my kayak so the attachment fit. The setup worked great. It held the camera up on the bow of the boat, and the pan head al-lowed me to smoothly swivel the camera if I needed to track a moving ob-ject. From then on, when I was in my boat, I was shooting video, a habit that eventually paid off.

I was the only continuous filmer in our search group. Brian was reluc-tant to cut his nice L.L. Bean kayak, and when his unblemished boat was stolen, he bought a new boat that had a very wide rim that made camera at-tachment seem impossible. We couldn't cut the boat that we borrowed from Mark to make a camera fit. When I suggested that I hack a slot on her new blue kayak, Wendy gave me a look that ended the conversation. So Brian and Tyler had their cameras in their laps or attached to their belts as they moved about the area. Kyle never carried a camera.

Our second video strategy was to film cavities and capture an image of an ivorybill coming or going. Our problem with cavity monitoring was that there were so many big cavities on the study site. It was easy for the search for a roost to begin to seem hopeless. I had to keep telling Brian and myself that we knew at least two ivorybills lived near our study area and that both members of the pair had to use a cavity each night. All we had to do was to get lucky once to hit the jackpot. We had to keep trying. In January we had about five Hi-8 video cameras that worked. I say "about" because although we had seven cameras, two or three were only marginally functional, and good ones kept getting wet when they were mounted overnight near cavities. Water usually knocked a camera out of action for a week or so until it dried. Brian kept busy setting video cameras each evening and retrieving them in the mornings. By the middle of March we were down to two functional cameras. Swamps are hell on electronic equipment.

Along with video, a major thrust of our search effort was to gather audio evidence. I've already described how we used seven listening stations, each of which collected 24 hours of recordings. Our listening stations gave us a means to remotely moni-

Me with my video camera mounted on the front of my kayak. (Photograph by Wendy R. Hood.)

tor much of the study site. We positioned the listening stations about 500 yards apart in a broad line from the south bank of Bruce Creek to Story Landing. As members of our search team began to detect ivorybills around the study area, it seemed that our placement of stations was pretty good. Wherever we detected an ivorybill, a listening station seemed to be nearby.

Originally I envisioned that Kyle would spend half his day swapping out batteries and memory cards and half of his day analyzing the recorded sounds. In this way, we would get feedback from the listening stations within a day or two, and our findings would direct Brian to the ivorybills. Such in-the-field analysis of audio recordings proved not to be feasible. Kyle spent essentially all day swapping batteries and memory cards at listening stations, and his laptop computer was constantly tied up just downloading and storing the audio recordings. All analysis had to be done in Dan's laboratory in Windsor, Canada, and that meant a least a two- or three-week delay in feedback from the stations. Given our limited funding and resources, there was no alternative.

Our second source of audio evidence of ivorybills was recordings made by searchers when they detected birds. The series of double knocks recorded by Tyler with his camcorder in December was a dramatic example. Such recordings were in many ways better than recordings from listening stations because they came with a context. Both Brian and Tyler had heard a bird double knocking right in front of them; Brian had heard a second more distant bird double knocking. They were certain that what they heard and captured with the video camera was not a distant automobile or someone bumping around in a boat. This sort of context made camera recordings all the more convincing. The microphones on the camcorders were not very good, however, so these recordings were not as high in quality as recordings made by listening stations with their Sennheiser microphones. They were adequate for clear and loud sounds produced by a bird that was not too far away.

Kyle had a Marantz recorder with a directional Sennheiser microphone, so he had the potential to make the best recordings of all. In the first six weeks of the search, a period in which Brian had more than a dozen ivorybill detections and I saw or heard ivorybills five times just on my weekend trips, Kyle detected no ivorybills. He seemed to be struggling to get up to speed on bird sounds and sights, and I think that, for at least the first couple of months, he lacked the confidence to identify the sound of almost any bird, let alone the sounds of an extremely rare bird like an ivorybill.

The potential was there, however, for Kyle to make the best recordings of the bird.

Earlier in the fall I had the chance to visit Dan and Stephanie in Windsor, Ontario, and to tour Dan's lab and see his setup for sound analysis. The University of Windsor is an urban commuter university that lies literally in the shadow of the Ambassador Bridge, just a couple of blocks from the beautiful Detroit River waterfront with the skyline of Detroit creating a picturesque backdrop. Detroit, originally called "les Etroits" or "the narrows," was founded by the French colonial governor Antoine de Cadillac because of its strategic location at the narrowest water barrier between what is now Michigan and Ontario. For millennia, migrating hawks that cannot traverse expanses of open water have used these narrows in the Detroit River as an escape route as they flee the approach of winter and move south toward warmer wintering areas. The movement of Sharp-shinned, Cooper's, and especially Broad-winged Hawks over this section of the Detroit River can be spectacular in late September and early October. The narrowness of the river makes downtown Detroit seem almost close enough to touch, even though it is in another country across the river.

Dan's research team occupies a spacious lab on the first floor of the rather old and decrepit Biology Building on Sunset Avenue. The first thing that I noticed when I entered the lab is that it was dimly lit. I had also noticed this about Dan's office at Auburn. His Auburn office had a big window that would flood the room with daylight if the blinds were pulled up, but Dan always worked under a small desk lamp in a dark room with the blinds shut tight. Now this lab was similarly shielded from daylight, and the overhead fluorescent lights were turned off. The room was only dimly illuminated by small desk lamps at each workstation and the glow from numerous computer screens. It gave his lab a very somber atmosphere and immediately imparted a feeling to visitors that very important work must be going on here.

I imagine that Dan has never thought much about lighting—it just sort of ended up dim. But I have to think that Dan's focus on sound and audio analysis rather than color and visual display (the specialty of his wife, Stephanie) led him to block out the distraction of the visual world while he works. In this way, he can focus that much more intently on sounds. Fast and efficient sound analysis was what we needed from Dan's group, so I was all for turning out the lights. I should point out that Dan lives in a darkened world only at work. Dan and Stephanie's house in Windsor is bright and colorful and cheery—no blackened windows (but maybe that's Stephanie's doing).

When most people, including most scientists, think about a research laboratory in a science department at a university, they think of a room with work benches cluttered with glassware and chemicals and analytical machines humming and spinning in the corners. One can imagine professors and grad students strolling about in white lab coats wearing goggles and rubber gloves as they measure and mix and record. This image is indeed how many labs in my department look, and even my own molecular ecology lab fits this description pretty well (although you'll never catch this professor in a white lab coat). But this is not a good description of most field ecology labs. Most ecologists run their experiments and collect their data in the field. The labs of most ecologists are used primarily for computer analysis of data, and most ecology labs resemble Internet coffeehouses.

Undergraduate students (left to right) Jason Mouland, Jessica Cuthbert, and Sarah Lippold scanning audio files in Dan Mennill's research lab at the University of Windsor. (Photograph by Daniel J. Mennill.)

A computer lab is a good description of Dan's research space at Windsor. The focus of Dan's research and that of his students is animal sounds, so Dan's lab was set up for complex audio analyses. Such audio analyses are done ex-

clusively with computers. Dan's lab was divided into half a dozen work-stations, each with a powerful CPU connected to twin flat-screen monitors. At each station sat a grad student or technician diligently analyzing sound files.

When Dan first explained his listening stations to me, I thought that the search for signature ivorybill sounds could be fully automated. In other words, I thought you could tutor a computer with known kent calls and double knocks and then push a button and have the computer efficiently find every kent call and double knock in an audio file. As seems to be true of just about everything associated with ivorybills, however, it isn't that simple. If double knocks and kent calls were given close to the microphones with little background noise, then, yes, a computer could be taught to find them. But ivorybill sounds seem to be invariably made at the very edge of the range of the microphones, and they are typically embedded in a stream of very distracting background noise. No current computer algorithm can efficiently locate such sounds in an audio file. There is no substitute for the judgment of a trained person.

Dan had funds for one full-time or three part-time technicians, and he elected to hire three students part time. It turns out that using three people part time was much better than one person full time. Searching the audio files from the Choctawhatchee River for ivorybill sounds takes tremendous concentration, and in the first few weeks of analyses, Dan realized that a person could only do such searching efficiently for a few hours per day. Beyond a few hours, even the most enthusiastic screener would start to get tired and slow down and make mistakes. With his three part-time people, however, Dan constantly rotated in fresh screeners and kept the effort efficient.

Dan recruited his part-time help from his ornithology class. He sought out the students with the higher marks in the class, but more than that, he sought out the students who showed the greatest curiosity about the natural world. He ended up with three undergrads, Sylvie Tremblay, Sarah Lippold, and Jason Mouland, who rapidly got very good at searching the audio files for kent calls and double knocks. Within a couple of weeks this team of part-time screeners was able to crank through about 72 hours of audio recordings per day. This rate of audio analysis was almost enough to keep up with what was coming in from the field, and if windy and rainy days were skipped, they could keep up. Such comprehensive file searches only lasted until Dan left to study Black-capped Chickadees in March. From mid-February on, searches of audio files became spotty, and when we were ready to publish our evidence for ivorybills in May 2006, less than half of our audio recordings had been searched.

Screening audio recordings from the swamp did not mean *listening* to all of the recordings. Rather, Dan's technicians searched the sound files visually. No matter how well trained a researcher's ears might be, humans accomplish better analyses of bird recordings if they can look at pictures of the sounds and search for characteristic shapes of target sounds. Dan's crew used a software program called Syrinx-PC in conjunction with a computer hooked to two screens. On the left screen they would scroll through graphical displays of recordings sent from the field. As they viewed the files, they would draw colored boxes around anything that looked interesting. The boxed section would be displayed magnified on the right screen. They could then compare these blown-up bits of sound to recordings of double knocks from all other *Campephilus* woodpeckers, to the sounds of Pileated Woodpeckers, to recordings made in the Singer Tract in 1935, to gunshots, and to Tyler's recording from his video camera. Syrinx-PC was an invaluable tool in our search for audio evidence.

The first audio project for Dan was digitizing the sound components of the two video recordings that Tyler and Brian had made in December. We were most interested in getting a digital version of the series of double knocks that Tyler recorded on Christmas Eve in Mennill Hammock. Actually, both Tyler and Brian had recorded the double knocks, but Brian didn't get his camera turned on as fast and was still fiddling with it when the double knocks stopped. Tyler's tape was by far the best. Dan sent me the digitized sound recording in early January. From within my e-mail I could play the sound clip. I put on large padded headphones, turned up the volume, and listened. Over the chatter of Carolina Chickadees and cheers of Red-bellied Woodpeckers, double knocks rang out: *bam-bam bam-bam BAM-bam BAM-bam BAM-bam BAM-bam BAM-bam bam-bam bam-bam.*

Nine double knocks were clearly audible. The middle five were the loudest and presumably the closest. The two before and two after that were much softer and seemed more distant. Neither Tyler nor Brian had heard the first two—probably because they were intent on getting their cameras running—but Brian had heard distant double knocks after the close bird stopped, and Brian was sure the last two knocks were by a different bird.

This is strong evidence for two birds, I thought. This recording did not

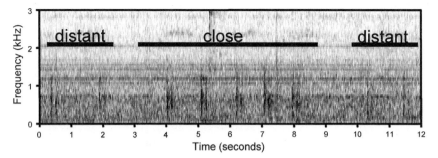

The sound spectrogram of the nine double knocks recorded on Christmas Day 2005 by Tyler with a video camera.

come from a listening station with no context. Brian and Tyler stood right in front of the bird that gave the louder double knocks. Brian and Tyler reported that there was no significant wind when they made the tape, and they were sure that the bird had not moved closer to them and then farther away.

Soon thereafter, Dan sent the audio file that Brian had made of the hammering ivorybill in Tit-ka Swamp. This sequence was less diagnostic than the series of double knocks, but as I listened to the recording I found it very intriguing. Mostly I heard Brian crunching noisily through dry leaves, but I could clearly hear the very heavy knocking of a woodpecker followed by a long pause where only Brian's crunching footsteps were audible, and then I heard "BAM," a single, very loud and powerful knock. I interpreted these sounds to be the bangs of an ivorybill that Brian was approaching on foot. The bird heard Brian coming, quit banging, and sat for a moment to see what was approaching it. When it saw Brian or thought the large thing approaching it had gotten close enough, it gave a single loud warning knock and flew off. In his book, Tanner wrote that ivorybills sometimes give single or double knocks when startled or alarmed. After the loud knock, Brian saw a large black woodpecker with white trailing edges to its wings fly off.

When Dan's group started to work through the audio files that I hauled out of the swamp and transmitted to them, there was a steep learning curve for the technicians for the first few weeks. Tyler and I were called on regularly to identify unknown sounds. One of the interesting things about the flooded forest along the Choctawhatchee River is that it supports a low diversity of birds in the winter. It is a very simple habitat—mostly big trees standing in water. The areas that stay dry all winter are small—apparently too small to support some locally common bird species. Lots of species are missing from the swamp that would certainly be present in the stand of pines at Bruce Creek Landing or along Highway 81. For instance, Song and Swamp Sparrows are abundant birds throughout the Florida Panhandle in the winter, but they are completely absent from our study site. The same is true for many other bird species.

Two of the most fortuitously missing bird species were the Blue Jay and the White-breasted Nuthatch. When members of the Laboratory of Ornithology discussed the challenges of identifying kent calls made by ivorybills in the sound recordings from Arkansas, they primarily mentioned Blue Jays and White-breasted Nuthatches as likely sources of kent-like calls. Blue Jays were thought to be particularly problematic because they had been recorded making calls almost identical to the kent calls of ivorybills. Tyler, Brian, and I, and for the brief time they were on site, Chet, Sidra, and Tyler's dad, Leon, all made a point to look for Blue Jays during our days on the study site. From December until late March we never recorded a Blue Jay on the study site, except one detection at the west edge of our site near listening station 2 in early March. Likewise, Dan's group never detected a Blue Jay vocalization in the audio files except on one occasion at listening station 2, the same spot as our March detection. Blue Jays were common in the adjacent uplands, but they rarely ventured into the swamp in the winter. I joked with Brian that Blue Jays are rarer than ivorybills in our study area during the winter.

Sound spectrograms of ivorybill kent calls recorded in the Singer Tract and two somewhat similar sounds, the chuck of a Gray Squirrel and the weep of a Wood Duck.

In the first few days as they were being trained, the technicians often mistook the plaintive weeps of Wood Ducks for a distantly recorded kent. Flagging Wood Duck notes as kent calls was a reasonable rookie mistake because on the spectrograph they look something like a kent call.

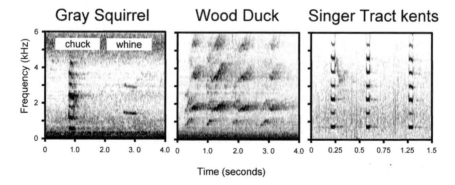

Tyler and I had no trouble identifying Wood Ducks, and very quickly the technicians learned the sounds of Wood Ducks and stopped flagging them.

Of all the sounds of the swamps along the Choctawhatchee, gray squirrels were consistently the major source of false kents. In the forests along the Choctawhatchee, gray squirrels are abundant, and they can make a lot of weird sounds. At a distance, some of these vocalizations could sound quite like kent notes. With practice, however, we learned that the kent-like portion of squirrel calls is usually followed by a drawn-out squeal, and squirrel vocalizations only sounded like kents when they were very distant and faint. Distant and faint kents we considered weak evidence under any circumstance. Once we filtered out all of the obvious false kents, we still ended up with an impressive 203 kents in the recordings made by the listening stations. Brian recorded seven additional kents with his video camera, giving us a grand total of 210 kent called recorded.

At first the technicians were overly sensitive to double knocks. Any two sounds that appeared in close succession would get flagged as possible double knocks. To give them a better idea of what an ivorybill double knock might sound like, Dan found examples of the double knocks of other *Campephilus* Woodpeckers—Pale-billed, Powerful, and Magellanic—on the web. With these sounds as a guide, the technicians came to understand that the double knocks of ivorybills should be bottom-heavy sounds and that they are easily distinguishable from two branches breaking in rapid succession, which was extremely rare *Spectrograms of two kent calls* anyway. Gunshots were not a problem for misidentifi- *recorded near the Choctaw-* cations. Although gunshots sometimes take on a double- *hatchee River that we attribute* knock quality, because they are so loud, the second bang *to Ivory-billed Woodpeckers and* typically has a long echo, making it easy to distinguish *a kent call of an ivorybill* them from sounds made by a bill hitting wood. *recorded in the Singer Tract* *in 1935.*

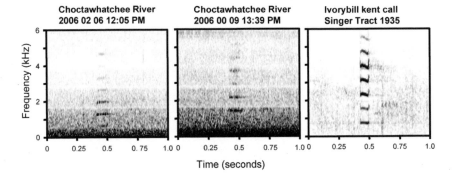

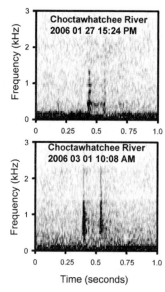

Frequency (kHz)

**Choctawhatchee River
2006 01 27 15:24 PM**

3

2

1

0

0   0.25   0.5   0. 75   1.0

Frequency (kHz)

**Choctawhatchee River
2006 03 01 10:08 AM**

3

2

1

0

0   0.25   0.5   0.75   1.0

Time (seconds)

*Sound spectrograms of two
double knocks recorded near
the Choctawhatchee River.*

Once they were trained and gained some experience, technicians more easily identified double knocks with certainty than they identified kents, and most of the double knocks that the technicians sent to us seemed like *Campephilus* double knocks to Tyler and me. We were always most impressed with multiple double knocks rather just a single double knock. In total, the listening stations recorded ninety double knocks, and Tyler recorded nine more with video camera.

As we proceeded through the winter and into spring, we kept hoping that we would get that elusive video of an ivorybill. We knew, however, that we might very well have to build our case for the presence of ivorybills along the Choctawhatchee River based on the dimensions of cavities, characteristics of scaled trees, and, most critically, sound recordings. Between the listening stations and video recorder sound captures, our goal was to accrue so much sound evidence in our small study area that it could not be dismissed. Without a video, we would not be able to call our results "definitive," "irrefutable," or "proof," but we could at least rise to the level of "compelling" or "highly suggestive" and justify getting enough grant money to conduct a follow-up search. Dan's sound lab was absolutely key to this plan, and I was extremely encouraged by what his group was compiling.

The Mule II

I'd like to be able to say that, from January to May, my primary contribution to the project was as coordinator, supervisor, and maybe even ivorybill hunter. Truth be told, though, my primary role during these months was to serve as a mule. Every other weekend I drove from my home in Auburn down to Florida and paddled my kayak into the site so that I could ferry digital audio recordings out of the swamp. When I got the audio data back to Auburn, I transmitted them to Canada so Dan's group could analyze them.

My first quick visit was to be Friday and Saturday, January 20 and 21, 2006. I had university and family obligations through Thursday evening, but I wanted a full two days to help with the search in the swamp. So on Friday morning I left Auburn at 3 A.M. and drove south in the dark. Instead of my usual route through Dothan, I took a more inspirational route through Enterprise and Prosperity, and shaved travel time down to just over 3 hours. I was paddling down Bruce Creek by 6:30, before sunrise. I didn't hear or see anything that suggested ivorybills that morning, but when I returned to camp around noon, Brian was waiting for me.

"Did you hear anything this morning?" he asked, obviously excited.

"No, nothing," I answered as I climbed out of my kayak and started gathering my cameras and GPS.

"I heard a bunch of double knocks," Brian blurted out.

"No way. How many is a bunch?" I asked, anxious to hear about this latest detection.

"At least fifteen," he answered excitedly as we walked the path toward camp. "I was sitting in Hill Swamp, pretty close to camp, quietly reviewing videotapes of cavities when I faintly heard a double knock. I wasn't even sure at first. I focused on the sound and then there it was again. And then again. And it was getting closer. Geoff, it was amazing."

"You thought the bird was moving closer to you?" I asked, wondering if he might have seen it.

"It never got very close. It seemed to move from north to south along the river. I think the bird was flying, landing to double knock, and then moving again. It seemed to be moving down Cougar Island toward Story Landing. From the first double knock to the last it seemed to move a long way—maybe more than a kilometer. It was hard to judge distance though."

"Did it just give one double knock each time before it moved?"

"No. Sometimes it was one, but sometimes it gave two or three then a pause, and then the next one was farther to my right."

"Did you record it?" I asked anxiously. I had been telling everyone to turn on their cameras as soon as they heard or saw anything interesting.

"No. I didn't turn my camera on. It seemed so faint. I didn't think the camera would record it. The listening stations might have gotten it, though. There are two stations that might be in range."

"That's a great detection," I said, putting aside my disappointment that Brian hadn't tried to record the knocks. "I really hope the listening stations picked that up."

"Yeah, it's been a very boring week. I've had nothing since Tyler and I had the kent calls last week. And then, a couple of hours after those double knocks, I came back to camp, sat down, and heard another double knock."

"Wow. From camp? I better stay alert," I said as I put my gear down on a chair next to my tent.

Beavertown, our main camp, sits on Beaver Hammock, a few acres of relatively high, dry ground that is completely surrounded by water when the river is up. One hundred meters to the north is the main channel of Bruce Creek. As this creek nears the Roaring Cutoff, it sends out two fingers of water that form the east and west boundaries of our hammock. The eastern boundary is a relatively wide channel with few trees and this is our front door, so to speak. This eastern channel is where we keep the canoe and where we usually keep our kayaks for the night. There is a well-worn path from the kayak pullout to camp. The western boundary channel, in contrast, is choked with tupelo and cypress. The water here tends to be shal-

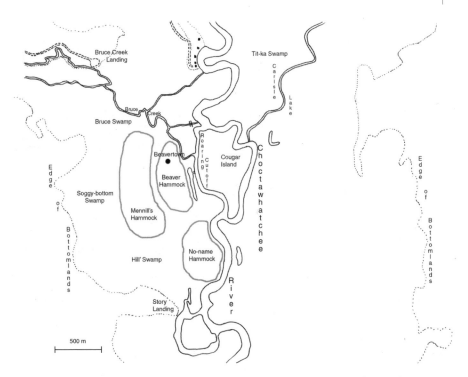

The following labels appear on the map:

Bruce Creek Landing
Tit-ka Swamp
Carlise Lake
Bruce Creek
Bruce Swamp
Roaring Cutoff
Beavertown
Cougar Island
Beaver Hammock
Choctawhatchee River
Edge of Bottomlands
Soggy-bottom Swamp
Mennill's Hammock
Hill' Swamp
No-name Hammock
Story Landing
Edge of Bottomlands
500 m

A map of our study area near the mouth of Bruce Creek showing the hammocks and swamps that we named.

low; during low water we can walk out the back of camp across this channel without even wearing boots. When the river is up, the water here comes to within 30 feet of Brian's tent. We use this back door to access Mennill Hammock and the swamps along the eastern side of Bruce Creek. These two water channels meet about a half mile to the south, and then one main water channel proceeds down to Story Landing. A faster way to Story Landing is to float down the Roaring Cutoff and the Choctawhatchee River. The current along this route is stiff, and we don't typically paddle up river. A day spent searching for ivorybills usually involved paddling up and down these channels. Sometimes we walk or wade around the hammocks, but one can move faster and more quietly in a boat, and that is generally what we do when water levels allow it.

The next morning I arose to watch a big oval cavity near the top of a relatively small, dead tree. The tree seemed just large enough to host an ivorybill-sized chamber within. This cavity was within 50 feet of the river, and I got to it by paddling down the river and struggling out of my kayak

The author paddling in the channel that served as the front door for the camp. (Photograph by Wendy R. Hood.)

on the steep bank, trying not to lose my boat to the swift current. Nobody was home in the cavity on that morning, so I gave up around 7:30 and decided to walk around in waders instead of kayaking. I intended to walk to another cavity that Chet found during his visit a couple of weeks before. From the description, Chet's cavity sounded like a cavity that I had found in November when I didn't have a GPS, and I wanted to find it to confirm that it was the same. Brian had marked the cavity on my GPS unit as cavity 22, and it was about a quarter mile south of the cavity I had just staked out. I was able to wade about two-thirds of the way to cavity 22 before the water got too deep and I had to turn back. I didn't want to paddle out in the current of the river, so I dragged my kayak through the forest for a few hundred feet until I found enough water to paddle in. I then started to float my way toward the cavity.

There is really no substitute for a kayak in this habitat. I was often floating across expansive areas covered with about 3 inches of water but also crossing channels that were 4 or more feet deep. Neither waders nor a canoe could have gotten me where I was headed. The kayak was perfect. And when I ran out of water, I could easily drag the boat along behind me.

As I got to within about 250 feet of the cavity, I was paddling through a deeply flooded stand of timber watching my GPS and trying to avoid the numerous obstacles in my path. Floating through a flooded woods is challenging because there is no clear path. You have to pick your way around the trees and over branches and bushes. Cypress knees always seemed to be lurking two inches beneath the surface of the water, so my kayak was constantly hanging up, forcing me to back up and go around. I was winding my way through the trees, often making a fair bit of noise, when suddenly a large bird took off from a tree near me. My main impression was black and white— mostly black but with bright white trailing edge on its wings. It was a naked-eye look, and the only color pat-

Mark Liu navigating his kayak through a flooded forest, which is challenging because there is no clear path. (Photograph by Geoffrey E. Hill.)

tern I noted was a broad band of white on the trailing edge of the bird's wings. I didn't notice any white on the back or neck and I didn't see bill color, but it was moving very quickly away from me.

It was an impressive flier. It seemed to cover a lot of air with shallow beats of its long wings. I estimate that the bird was 30–50 feet from my boat (I moved my boat before I thought to measure the distance) and maybe 3–5 feet above the water when it took off. It was already flying when I saw it.

I had no chance to raise my binoculars. If I had been walking on a trail with binoculars in my hands, I probably would have raised them. But I was in a kayak with a paddle in my hands, so I just watched the bird. It flew powerfully, rising rapidly up toward a distant tree line. It was in view for about 4 seconds, by which time it was 300–400 feet away into distant trees. I could see it braking to land on a distant trunk, but I never actually saw it land. It was definitely landing like a woodpecker on a vertical trunk and not on a branch like a crow. Just as it was about to land, it disappeared from view behind vegetation.

I reached down, turned on my video camera and placed it in my lap. Just then I heard a single knock and then a double knock to my right. It sounded pretty close, but distance was hard to judge. A dense tangle of bushes was about 70 feet to my right and it was certainly behind that, so 100–200 feet away is my best guess. In retrospect, I probably should have paddled immediately toward the bird that appeared to land in the distance, but charging toward a bird in a kayak in a flooded forest is not like charging toward a bird across an open field. A straight route was impossible. Any approach by boat would be noisy, and I would have had to focus more on maneuvering the kayak than on looking for the ivorybill. Instead I just sat and watched, hoping the distant bird would give me another look or that the second bird would move across my field of view. Neither happened. I sat and watched for about a half hour, paddled through the area where the bird had landed, then headed back to camp.

On my way back to camp, I came upon a Pileated Woodpecker banging on a tree. For a moment after I heard the first bangs I was hopeful that it might be an ivorybill, but after listening for a few seconds it was clear that the bangs were not powerful enough and were too fast. I carefully approached the bird, but it still flushed at about 100 feet and flew away from me. It was a very useful look because I could compare this fleeing pileated to the other bird I had just seen in the same light and viewing conditions. The pileated really didn't look much like the bird I had just flushed. The wings of the pileated were more rounded and its flight much less direct and powerful. The pileated flashed a lot of white on its wings, but the white was at the base of the primaries, not on the trailing edge of the wings. Also, the red on the pileated was conspicuous—very hard to miss in the gray and brown swamp. I had seen no trace of red on the bird I had just watched. Failure to see red may have been due to the naked-eye view and distance— I had failed to see dorsal stripes or bill color, and I attribute that to viewing conditions—but after seeing how conspicuous the red on the pileated was, I doubted I could have missed a red crest on the ivorybill if it had been present.

It turns out that paddling away from the location of my ivorybill encounter after 30 minutes was a mistake. The ivorybills came back. Between 2:19 and 3:45 that afternoon, listening stations near the spot where I flushed the ivorybill recorded two double knocks. Throughout this first field season we had the mistaken idea that if an ivorybill was flushed from a spot, then it was gone from that spot for good. We made no effort to stake out specific locations where we encountered ivorybills. Our listening station data, however, indicated that ivorybills hung out in relatively small areas for days at a time and apparently returned to such favored spots even after they encountered a person. Unfortunately, Dan didn't discover this site tenacity until well after our spring search was over.

Not only were ivorybills back in the vicinity of my encounter later that day, but listening stations indicated they were at the same location on the two days before, with two and ten double knocks recorded on these days. Ivorybills were also back on the day after my encounter, with five double knocks and three kent calls recorded. Looking back, we missed a great opportunity to get in position and wait for an ivorybill to come to us. If we had staked out the area in a blind or full camo, an ivorybill may have flown within camera range. Regrets aside, I really like these multiple layers of evidence. Skeptics could reasonably discount my sight record because I failed to see details on the bird such as a pale bill or dorsal stripes, although I'm a very experienced birder, and I'm sure of what I saw. But what are the chances that I misidentified a pileated woodpecker at close range just as I heard a nearby double knock? And then what are the chances that a few hours later the listening stations closest to this encounter would record kents and double knocks? And finally, what is the probability that only these nearest listening stations would also record ivorybill sounds on the days preceding and following my encounter? I might also note that the link between what I experienced and what the remote listening stations recorded was made blindly. The students searching audio files did not know about my encounter until after they had flagged the kents and double knocks, and I didn't know about the corroboration of my sighting on the audio recordings until months later.

I was fighting a cold that week, and I drove back to Auburn that evening feeling ill. My fever couldn't keep my spirits down, though. Our ivorybill search was going very well. Ten days worth of audio recordings from the listening stations were stowed in the hard drive on the seat beside me. I had finally gotten a good look at an Ivory-billed Woodpecker—the last of our trio of discoverers to do so. We were detecting ivorybills at a rate that had to be orders of magnitude greater than what was happening in Arkansas, and slowly we were gathering tangible evidence—all audio to this point, but audio evidence might be sufficient if we got enough good recordings.

On Sunday night, the day after I drove home, as I was helping my kids with their homework, the phone rang. It was Brian. I figured the only reason he would be calling so soon after my visit would be with good news—maybe he had captured an image of an ivorybill. Unfortunately, my little daydream about getting word of a definitive video was short lived.

"Geoff, I'm in trouble," Brian said as soon as I answered the phone. He was obviously worked up. "I mean I'm all right. I'm not hurt or anything. But my boat's gone."

"What do you mean your boat's gone?" I asked, suddenly getting worked up myself.

"I paddled up to Bruce Creek Landing and drove to town to get some supplies," Brian explained. "I forgot my lock back in camp. I hid my boat in the bushes but I guess not good enough. Someone took it."

"Oh no. Not your nice kayak," I said, feeling Brian's loss. He had shopped around for weeks at the end of the previous summer and had found a perfect kayak, just the right size with all the right features for swamp work. In particular, Brian had gotten a boat with a comfortable, adjustable seat. Although he was only twenty-five, Brian had already spent years helping his dad lay flooring in Pennsylvania. His back tended to give him trouble if he spent long hours in a kayak with a poor seat. My little green kayak, for instance, was legendary for its level of discomfort. It has a nonadjustable, hard plastic seat, and most people couldn't sit in it for more than an hour without complaining.

"Can you contact Kyle and tell him?" I asked, realizing that the biggest immediate problem was having Kyle in camp not knowing what was going on.

"I don't think so. His cell phone doesn't work from camp. I didn't bring a radio with me." Brian and Kyle had both gotten very sloppy about carrying two-way radios with them. When Brian calmed down and we solved this problem, I meant to use this little episode to try to inspire them to keep in better radio contact.

"That's too bad. Kyle is going to think you drowned in Bruce Creek," I lamented.

"Hey—you know that old railroad bed that leads to the mouth of Bruce Creek, I could park on the road and hike to the mouth of Bruce Creek. It wouldn't be that far."

"But you'd be on the wrong side of Bruce Creek," I stammered, not liking where Brian seemed to be headed with his idea.

"I could take off my clothes, put them in my dry bag, and swim across the creek," Brian continued, not sounding all that convinced himself.

"Brian, you're crazy," I said. "That water is freezing. You'll drown for sure. Do not try to swim across that creek."

"OK," Brian relented, knowing it was a desperate idea, "but what do I do? Without a boat, I can't get back to camp." He was starting to calm down and rationally assess the situation.

"You have to have a boat to continue the study, and we're sure not bailing on this project because of one lost boat."

"Stolen boat," Brian corrected.

"Yeah, yeah, stolen boat," I continued. "Kyle will just have to think you're dead for a day. Start driving to Tallahassee."

"But that's 100 miles away," Brian protested weakly as he began to realize the plan.

"107 miles," I corrected. I had moved over to my computer while we were talking and had just gotten the exact distance from Ponce de Leon to Tallahassee from Mapquest.

"Tallahassee's a big city. I know you can buy a new kayak there. Start driving and I'll find a place. Maybe you can still get one tonight."

"All right," Brian agreed, "but first I'm going to try to find Earl. Maybe he saw who took my boat."

"OK. I'll call back in a few minutes when I find a place that sells kayaks."

The best place I could locate in Tallahassee with my Internet search was the Sports Authority. There was a store on the west side of town near the expressway. They were still open when I called them, but they were due to close in 30 minutes—no way Brian could make it. They had a couple of kayaks in stock—I forgot to ask what color. They opened at 10 the next morning, but they were on Eastern Time, so Brian could get a new boat by 9 A.M. Beavertown time. I called Brian back, gave him the address, told him to stay in a motel that night and get back in the swamp as soon as possible the next morning.

It turned out the only kayaks that the Sports Authority had in suitable models were either canary yellow or fire engine red. Brian bought the red one. Dan had left three cans of green spray paint in camp from his construction of listening stations, and Brian set to work right away "fixing" the red kayak. Three cans of spray paint later he had a hideously ugly red kayak blotched everywhere with green. It was way more cryptic than a solid red kayak, though, which was critical. Brian later told me: "When I paddled that bright red kayak down Bruce Creek, it seemed to upset the animals."

"Really? You could actually notice a difference?" I asked, a bit surprised. I just figured that the difference between a green and red kayak would be a few feet closer approach to an ivorybill. I didn't think you'd instantly notice a difference in how small birds reacted.

"Yeah. The chickadees and titmice were actually scolding me as I paddled down Bruce Creek, and I seemed to be surrounded by small birds. I think they came close to see what could be so big and red," Brian replied. Then he added with a smirk, "Maybe I should have kept it red. Ivorybills might have dive-bombed my boat."

"Too late. It's now the beautiful color of sludge."

We were pretty sure that no one would want to steal this hideous monster. Brian was back in action before lunch on Monday, and Kyle, I heard later, was quite relieved and a bit surprised to see a living Brian come walking back into camp.

Over the next few weeks, sightings continued to mount. Most encouraging, we continued to detect birds in pairs. Brian called me on Febuary 1 to tell me he had his best look yet at an ivorybill.

"What did you see?" I asked.

"It had been another slow week. I hadn't seen or heard a thing in days. Then I was hiking through No-name Hammock near the river . . ."

"I thought I had renamed that Brian's Hammock?" I asked, knowing that Brian was too modest to name a landmark after himself.

"OK, Brian's Hammock—I just got used to calling it No-name Hammock. Anyway, I was quietly walking through the hammock when I flushed two birds from a tree."

"Did you see them perched?" I asked. Despite many sightings, no one in our group had yet seen a perched ivorybill, and I was eager for someone to get a clear look at the white triangle on a perched bird.

"No, they had just taken off when I saw them. It was kind of funny. I think I really startled them, and they seemed confused for a few seconds. They kind of flew out and hesitated before one took the lead and headed for the river and the other followed."

"Did you see them well?" I asked, wanting to know the field marks he had seen.

"Oh, yeah. It was my best look. They were pretty close, less than 200 feet away when I flushed them, and as they wheeled and hesitated I saw the underwings of both birds clearly. They were white on the forewing and trailing edge with a black band down the middle of the wing. I could see the wing pattern very clearly with my naked eye."

"What about bill color or a crest?"

"No, I didn't see bill color and I didn't see red on either bird."

"Hmm. That close, you'd think that you would have seen red if one had been an adult male," I mused.

"Yeah, it seemed to me that I should have seen red, too. I didn't though. They flew off and out of sight toward the river. The river was only a few hundred feet away, but I have no idea if they turned before they reached it. I could only see them for maybe 200 feet as they flew off through the woods."

I said good-bye to Brian and hung up the phone. No doubt Brian had seen two ivorybills together, but why hadn't he seen red on one of the birds? It seemed as if they had to be a pair—but maybe not. Maybe it was a female and her daughter from the previous breeding season. Tanner had noted that young sometimes stayed in their parent's home range through their first winter. But this close to the breeding season, pairs should be together much of the time. Maybe Brian just missed the red on the male. It would take more sightings to figure this out. I was starting to think that it might be harder to see red on a flying ivorybill than on a flying pileated.

My next visit to the study site started on a foggy Friday morning on

Febuary 3. I spent the first hour of daylight around our listening station 2, not far from camp, where Dan had recorded a bird banging on a tree for several hours a few days before. Birds were quiet in the fog, and I couldn't find listening station 2 (I didn't have the GPS coordinates), so I headed for Story Landing. I had a nice morning paddling downstream below Story Landing as the sun slowly burned through the dense fog. I hadn't realized that the old river channel forms a nice oxbow lake here. On our topo maps, this oxbow is still marked as the main river channel, but that is no longer the case. Almost all the water in the river goes through what is marked on the map as a small cutoff, and this wide loop in the river is now essentially a lake with a beautiful forest of cypress and oak bordering it.

At the north end of this oxbow is Story Landing and the start of the cypress stand that runs almost to camp. I got out of my kayak at Story Landing to look around a bit. Tyler had given me a tour of this spot back in January. Where the old road comes down to the water—the actual abandoned landing—there is an old, dilapidated fish-cleaning station. The forest here is impressive, with scattered huge oaks and spruce pine forming an open grove. Unfortunately, the boundary for the land owned by the state of Florida is only about 200 feet above the landing. The beautiful forest continues up the hill, but there is a new gate across the road, a new fence along the property line, and there is a newly cleared or expanded road through the forest on the private land. Relatively few trees were felled in making this clearing, and the road is just a dirt path, but it looks like the area is being readied for development. Ivorybill feeding trees were scattered all over this area, including some of the trees that had been pushed aside when the road was cleared.

Later in the day, Brian and I discussed the Story Landing area.

"Do you think that land can be preserved for ivorybills?" Brian asked.

"If we can convince the world that we've found ivorybills, then I think the state of Florida, Feds, or the Nature Conservancy will want to buy nice parcels of land like the area above Story Landing. That would create a buffer from development or disturbance for the ivorybills we've been monitoring. Who knows, this landowner might turn the property into a private ivorybill preserve and make money leading tours onto the site. One way or another, I'm confident that the forests around Story Landing will be preserved once we announce ivorybills."

"You really think our discovery will lead to preservation of some of this area?" Brian asked.

A gate marking the edge of state land at Story Landing with a recently cleared lane. (Photograph by Geoffrey E. Hill.)

"It should. They are buying thousands of acres in Arkansas because one bird might be in the area. If we have good evidence for a breeding pair or especially a population of ivorybills, I would think that a huge effort would be made to preserve all the swamp forest along the Choctawhatchee."

I took a few photos of the gate and clearing, but I didn't dally long. It was too depressing to see a beautiful hammock with ivorybill feeding trees being disturbed. On my way back to my boat, about halfway between the gate and the landing I noticed a big cavity that had ivorybill potential. It was in a dead sweetgum, and there were bark chisels on the lower portion of the trunk. Somehow Tyler and I had missed this on our earlier visits.

By now the fog had lifted and it was a bright day. The sun wasn't out yet, but it was just barely obscured by a brilliant haze. Bird activity had picked up as I paddled north into the cypress stand and headed toward camp a half mile away. As I entered the trees, I heard slow and powerful banging. It sounded like an ivorybill to me. I made sure my new digital video camera was on. I had just bought this camera so I'd have a means to record birds as I went back and forth from camp; we had too few of the old cameras for me to keep one all week in Auburn. I was in the midst of a feeding flock of passerines that seemed to be making up for foraging time lost to fog that morning. Yellow-rumped Warblers, Ruby-crowned Kinglets, and Blue-headed Vireos flitted back and forth through the trees, distracting me as I looked for the banger. A Red-bellied Woodpecker fluttered past, going from right to left, and then a bit farther away a Pileated Woodpecker undulated past, flying left to right. The pileated flew from near where I had heard the banging—was it just a pileated I had heard? No. The banging continued. As I pointed my camera in that direction and slowly paddled the boat forward with one hand, a large, long-winged bird flashing lots of white in its wings flew through the trees in front of me. It was a poor, fleeting look through the trees. Later, as we reviewed my videotape, I could be heard exclaiming to myself, "Now that was interesting."

I had the video camera pointed right at the bird as it flew past, and I thought maybe I had just captured an image of an ivorybill. Unfortunately, my view of the bird had been against brilliant, hazy sky. The video that I shot was terribly backlit and the camera focused on trees about 50 feet away. The bird had been at least twice that far away. There wasn't any trace of a fleeing bird on the video.

"You didn't think it was going to be that easy, did you?" Brian asked later as we reviewed the tape on his laptop computer and found nothing.

"Yeah, actually I did. Every time I get into this swamp I expect to get a video of an ivorybill. I guarantee one of us will get a good video very soon."

"If I wasn't so broke I might take that bet. I've been chasing these birds for almost two months now, and I'll tell you they won't let you just walk up and take their picture."

"Maybe they'll get more distracted when they start breeding," I mused.

Voices in the Wilderness   **12**

A s our search entered February, Brian was feeling overwhelmed. It was clear that we had at least one pair of ivorybills using the area between Bruce Creek and Story Landing. We were repeatedly seeing and hearing two ivorybills in that area; there were many fresh cavities and feeding trees; and our audio evidence was mounting. By this point Brian had conducted more than 40 cavity watches by video or in person with no hint of ivory-bills coming or going, and neither he nor I had come close to capturing a bird on video as we paddled around. Kyle had fallen into a routine of main-taining the listening stations, marching or paddling along the same routes with his head down, and then spending chunks of the day in his tent man-aging the enormous computer files that the listening stations generated. Kyle never got out of his tent in the morning until long past sunrise, and he didn't carry a video camera with him—he said he had too much equip-ment as it was. Dan and I were pleased that Kyle was managing to keep all seven stations operational, but having Kyle totally committed to the sound work and apparently disinterested in looking for ivorybills meant that Brian was the only ivorybill hunter in the swamp when I wasn't there. If we were going to get a video of an ivorybill, it would have to be by Brian or me.

By February, the lack of kent calls was bothering all of us. Without ex-ception, every biologist who had previously observed a breeding population of ivorybills commented on how they heard nasal kent calls of ivorybills as they moved through the forest. Since December we had regularly heard and recorded double knocks, but members of our crew had detected only a few kent calls. In his book *In Search of the Ivory-billed Woodpecker,* Jerry Jackson quotes passages from John James Audubon's trip down the Mississippi river in the fall of 1820. In November and December Audubon frequently heard what he called the "paits" of ivorybills, suggesting that in many areas they were the dominant sound of the forest—more frequently heard than the whinny-like calls of pileateds. Why were we detecting no kents?

The dam broke on February 7, 2006. Brian was quietly walking on the south side of Mennill Hammock near a section of tornado damage when kents rang out. He missed one kent as he got his video camera started, but he managed to get recordings of seven kents calls—five distinct on the video recording and two very faint. Then the ivorybill got quiet for about 10 minutes, and Brian turned off his camera. Almost immediately the kents started again. He heard four more kents before he could get his camera on, and then they stopped—for the rest of the day as it turned out. Brian was ecstatic. If birds started giving kents, they would be much easier to find and track, and our chances of finding a roost and a nest and getting video of the birds would increase tremendously.

The next day Brian was out all morning in the same area he had heard kents the day before, but he heard nothing that sounded like an ivorybill. He and Kyle planned to head to town around lunch because Brian had to send me some paperwork for his grad school application. When Brian got back to camp, Kyle mentioned rather off-handedly that he had heard some weird sounds from camp that morning. Brian didn't think that much of it because Kyle was still learning the sounds of the forest, and lots of birds were starting to sing and make different sounds as the days got longer and warmer. They loaded their canoe at the channel below camp and started to paddle north up the Roaring Cutoff toward the mouth of Bruce Creek when Brian heard kents. The kent calls were coming from two birds—one somewhere behind camp near Mennill Hammock and one up Bruce Creek.

"Kent calls!" Brian exclaimed. "Let's try to find the birds." And he and Kyle started to paddle hard up the roaring cutoff and then up Bruce Creek. Brian turned on his video camera and put it in his lap thinking it was recording the kents as they paddled. It didn't. When Brian played the tape for me later, I called it the worst recording ever made. The video was nothing but a closeup of Brian's jacket, and the audio was a horrible raking sound as the camera microphone slid back and forth in his lap. As they turned up Bruce Creek from the Roaring Cutoff and closed in on the kents, the birds became silent. Brian and Kyle stopped paddling and sat quietly and listened when they reached the spot where the kents had come from, but there was no hint of ivorybills.

As they sat drifting in Bruce Creek hoping an ivorybill would appear, Kyle looked at Brian.

"That's what I heard from camp this morning!" he exclaimed.

"Really? Kent calls? You're sure?" Brian responded, wishing he'd asked more questions about the mystery calls. The paddle they had just completed, up the Roaring Cutoff and lower Bruce Creek, circled camp. They were never more than a few hundred feet away. It made sense that if ivory-bills were in the area giving kent calls, Kyle would have heard them from camp.

"I'm positive. I heard a lot of those calls," Kyle responded firmly.

"How many?"

"I didn't count each one, but I'm sure it was more than fifty. It went on for a long time. I didn't know that's was what a kent call sounded like," Kyle answered, a bit embarrassed that he hadn't recognized kent calls. Then as an afterthought he added, "I did record the sounds, though."

"With your good microphone?" Brian asked, glad to have something to focus on other than the fact that Kyle had never even bothered to learn what a kent call sounded like.

"Yeah. After I listened to the strange sounds for a while, I got out the shotgun mike and recorder. Unfortunately, one of the mikes wasn't working. It had water in it or something. By the time I found another mike, swapped it out, and started recording I only caught the last kent or two just before the sounds stopped."

"Out of fifty kents you only recorded one or two?" Brian asked, clearly disappointed that neither of them had gotten much in the way of record-ings after two days of kents. "Maybe ivorybills will start kenting more often," he added hopefully.

The dream that kents would become routine never materialized. As a matter of fact, the day after Kyle and Brian heard the kents near camp, a storm front passed through the area, bringing the coldest weather of the winter, and the ivorybills became completely silent. Brian didn't detect an ivorybill for the next eleven days, the longest period with no detections of the entire winter and spring.

Three weeks later I brought my friend Dave Carr along when I came down for a weekend in the swamp. I was anxious to show Dave the site be-

*David Carr just before he heard
a kent call and double knock
along Bruce Creek.
(Photograph by Geoffrey E. Hill.)*

cause he was skeptical of our claims of ivorybills. We'd been birding together since we first identified the birds that we shot with BB guns as kids. He wouldn't hesitate to question one of my bird reports. He didn't come right out and say that he thought we were delusional about ivorybills, but he kept asking hard questions about why no one in our search crew ever saw an ivorybill perched or got a photo or found a roost. When I would talk about him getting Ivory-billed Woodpecker on his life list, he would get very quiet—he clearly was thinking that there was no chance that our fictitious ivorybills would end up on his bird list. I wanted to see his reaction to the cavities and feeding trees on the site, and I was really hoping that he'd hear or see an ivorybill.

Dave and I got to Bruce Creek Landing early in the morning of Friday, March 3, 2006. The water was high on the weekend of Dave's visit, and I wanted to explore the area west of camp on our way in. The search crew had spent little time in this area, but it was adjacent to areas where we had many recent ivorybill detections, and the previous week I had found two huge, fresh cavities in a young, solid cypress. Dave and I paddled around for the first 3 hours after daybreak in what seemed like endless tupelo/ cypress swamps. We found a few old cavities and some old feeding trees, but no fresh sign of ivorybills. Around 10 A.M. we found ourselves somewhere in the endless expanse of Soggy Bottom Swamp behind Mennill Hammock. On my GPS, I could see that listening station 4 was a few hundred feet to the east of us when I started hearing faint kents. I'm leery of calling distant kents. So many things can sound like kents when they are very far off. The more I listened to these sounds, though, the more convinced I became. It was two birds somewhere off to the east. The kents were well spaced, maybe 20 seconds apart. Dave couldn't hear them. I heard about twelve kents in all, about six from each bird. There was no way to charge after the sounds. The birds were far away, and I knew that moving in that direction would mean repeatedly crossing dry spots and deep water channels. My waders were down in camp.

Late in the afternoon, we walked around Mennill Hammock. We found the broken base of a small tree that had been hammered on a bit. The most interesting thing about this tree was that there was a spot where the bird had smacked a piece of bark and caught the hard wood underneath with its bill. A paper-thin section of wood had been curled, as if it had been struck by a sharp blade. Dave commented that such a curled section of wood suggested that the bird that had cut it had a flat, chisel-like bill. The next afternoon, we went back past the little tree that had the curled wood the day before and noticed that it was now whacked to pieces, with a large pile of wood shards lying underneath. I pointed out the shredded tree to Dave and suggested that an ivorybill had come back to this tree within the previous 24 hours.

"You know, we could net these birds," I said to Dave as we stared at the ruins of the little tree.

"What? Are you crazy? You can't even get a good look at them or a single picture and you think you can catch one?" Dave responded incredulously, looking at me like I might belong in an institution.

"These ivorybills fly up and down this road bed very low. They should be relatively easy to net. It just might take a few days or weeks," I continued, unfazed by Dave's skepticism.

"Geoff, they'd fly right through your nets," Dave commented.

I was pleased that I had already moved Dave's thinking from "only a lunatic would think he could net an ivorybill" to "the net wouldn't hold them." Dave had started out his academic career as a herpetologist, contin-

A small hardwood near Beaver-town camp with a curled sliver of wood that looked as if it had been struck by a sharp blade. (Photograph by Geoffrey E. Hill.)

uing the salamander- and snake-chasing that we had begun as kids. In graduate school he switched to botany. Birding was mostly a hobby for Dave, although he had directed some undergraduate studies of songbirds and had some experience netting them.

"Yeah, through my House Finch nets," I agreed. "We'd use more heavy-duty nets like they use at owl- and hawk-banding stations. Nets made for Cooper's Hawks and Boreal Owls would snag an ivorybill just fine."

"You're serious," Dave said, still incredulous. "You think you can net an Ivory-billed Woodpecker."

"Yeah, I really do. But not this year. We would definitely not try that without permission from the USFWS, and we'd need a much bigger crew. I wouldn't leave a net set for ivorybills unattended longer than 30 minutes ever. I can't imagine anything worse than capturing an ivorybill and hurting it. We'd also have to have a good reason to try to net an ivorybill like radio tracking an individual to determine the size of its home range."

We ended the day silently, drifting the last 500 yards of Bruce Creek toward camp. Right before sunset a kent sounded from the woods to the south. I raised my arm and pointed in the direction of the kent, and Dave nodded to indicate that he had heard it, quietly raised his camera, and turned it on. I already had the camera mounted on the front of my boat running. Sadly, a check of the videotape later turned up no kent. About 10 minutes later in fading light, I paddled over to Dave.

"What did you think of that kent?" I asked.

"It sure sounded like the kents on the Cornell website. The double knock was maybe even more impressive, though," Dave responded.

"What double knock?" I asked. The kent had been loud and clear, but I hadn't heard anything interesting in the subsequent 10 minutes.

"About 2 or 3 minutes after the kent there was a double knock. Didn't you hear it?" Dave replied, clearly pleased that he had finally heard something that I had missed. All day I had been hearing distant or high-pitched sounds that Dave couldn't hear, including the earlier kents. But I hadn't heard the double knock.

"Are you sure?" I asked, suddenly playing the skeptic. We had been floating within 30 feet of each other in the quiet evening. I couldn't imagine how I could have missed a double knock.

"It sounded clear to me," Dave added, not sure what else to say.

I was doubtful because I didn't want to admit I'd failed to hear a double knock, but later when we reviewed Dave's tape we could all hear it.

I couldn't find the double knock on my tape. There was one point, however, when the current moved my boat into some submerged branches that made a fairly loud noise as they raked along the haul. The branch scraping must have obscured the double knock.

When we talked about what he had seen that weekend, Dave had to admit that what he saw and heard was very suggestive.

"This swamp is amazing. There are so many big trees and endless sloughs and channels. It seems like there is room for ivorybills here," Dave said.

"What about the ivorybill sign that seems to be everywhere?" I asked.

"I'm not sure what to make of the feeding trees. I don't know enough about how pileateds feed, and I've never paid that much attention. I'll start to look more closely at Blandy [the field station where he worked and spent all of his time]. There are lots of pileateds there," Dave concluded.

"Take photos of the feeding trees that you find at Blandy so we can compare them to the feeding trees in this swamp," I suggested.

"Okay, I'd be happy to do that."

"What about the cavities?" I continued. I always thought that the cavities were hard to explain if there weren't ivorybills living in this swamp.

"The cavities do seem big, and there are lots of them. I've never seen so many cavities in a forest before. But again, I'm not sure that they are different from pileated cavities," Dave reflected. "I have summer research students who need projects. Maybe I can get them to find and measure pileated cavities at Blandy."

"OK," I continued. "So you're reserving judgment about feeding trees and cavities; what about the sounds you heard?"

"The sounds are by far the most exciting thing. I'm really amazed that I heard double knocks and kent calls. I didn't think I'd hear anything like that," Dave replied.

"And a double knock and a kent call from the same spot within a few minutes of each other," I added.

"Yeah, that's pretty damn convincing. It seems almost impossible that we'd hear both kents and double knocks coming from the same spot in the forest if there wasn't an ivorybill. I could imagine a pileated or something giving a very atypical drum that sounded like a double knock, although I've never heard that before. And I could imagine a Blue Jay or something sounding like a kent, although I haven't heard a Blue Jay since we got into this swamp. But what are the chances that both would happen at the same spot at the same time? I agree. I think we heard an ivorybill this evening."

"So Ivory-billed Woodpecker is on your life list now?" I asked, referring to the list of birds seen in North America that most American birders like to keep.

"No. I don't count H's," Dave answered seriously. ("H" was our shorthand for birds heard but not seen. We started calling such birds H's because that's how we'd code them on a checklist.) "Ivorybill and Flamulated Owls will be on my H list, but I won't count them toward my ABA total."

Even if ivorybill had to remain on Dave's H list, I felt very good that I had brought a skeptic into our swamp and convinced him that ivorybills still dwelled here.

After mid-February the research team continued to hear kents occasionally, about as often as we heard double knocks, and the listening stations began to record some kent calls. We knew it was time for ivorybills to be nesting, and I kept waiting for Brian to call me and tell me that a nest had been found. I waited and waited. No nest.

In late February I found a very fresh cavity in a solid young cypress that was about 18 inches in diameter. This fresh cavity was about 3 feet above a slightly older cavity, and both were huge and oval, cut right in the trunk of the healthy tree. I thought that one of these cavities had to be an ivorybill nest. But after days of watching these cavities with no activity, we felt foiled again. Then the situation at these cavities started to get very weird. One morning when Brian visited the cavity tree, he saw blood on the trunk above the lower cavity and a few feathers stuck to the bark of the tree. Thinking that a predator might have gotten ivorybill chicks, he used his climbing tree stand to climb up to the cavities the next day. The "blood" turned out to be red cypress sap. The feathers were large, white down feathers, which temporarily gave Brian some hope that they might be from an ivorybill chick. He reached as far as he could into both cavities, but he couldn't touch the bottom on either.

I've spent my career studying passerine birds, not woodpeckers, so when Brian told me he had white down feathers from the cavity, I also was hopeful that they might be from ivorybills. Then I read that woodpecker chicks lack downy feathers, so they couldn't be from an ivorybill. Something else must have nested in that tree.

Two days after Brian climbed the tree, I was on site and paddled past the cypress cavities. I was stunned to see a large rat snake about 2 feet below the lower cavity. These cavities were 30 feet up on the trunk of a smooth, branchless cypress standing in water. That snake had swum to the tree and climbed the smooth, vertical trunk. Not only that, but the snake had clearly eaten something about the size of a rat—it had a conspicuous bulge in its body. I watched the snake struggle into a smaller and older cavity below the two large fresh cavities. If I could have caught it, I would have killed it and cut it open to see what it had eaten. What a way to document ivorybills— cut from the belly of a snake. It squeezed into the cavity and disappeared, though. We'll never know what it ate, but I doubt very much that it was ivorybill chicks. Ivorybill parents could have whacked a snake like that to pieces and would never have allowed it so close to their nest.

In early March, Brian spent a weekend with his girlfriend in Panama City. It was his first break since December. David Carr and I were around for the first two days that Brian was gone, but we left Sunday morning and Kyle was by himself in the swamp for the next two days. On Monday, March 6, with Brian due back that evening, Kyle was making his morning rounds swapping out batteries and memory cards at the listening stations. The river was up, so he was mostly working from his kayak. He had just swapped out the Beaver Hammock, Mennill's Hammock, and Bruce Creek stations and was headed down the river toward the three listening stations

A large rat snake, stomach bulging from a recent meal, going into an old cavity a few feet below two larger and fresher cavities on a cypress. (Photograph by Geoffrey E. Hill.)

on the south side of the study site. As usual, Kyle was paddling with his feet up on the bow of the boat. The kayak was too cramped to easily accommodate his long legs, and on this morning he was wearing waders with big clunky boots, making it even tougher to jam his legs inside. It was much easier to just stretch his legs out on the bow. Unfortunately, riding with his legs on top made Kyle rather unstable in his boat. But this was just a routine trip, the same route Kyle had taken dozens of times.

As he moved down the swollen river, Kyle's kayak suddenly slammed into a submerged tree. His boat slid up onto the hidden trunk, tipping him perilously to one side. As he struggled to push the boat off the snag and keep his balance, the current whipped his boat around perpendicular to the flow. And then he was in the cold, brown water, flailing and trying to find the surface. His first thought was to save the precious batteries and memory cards that he carried, but within an instant those worries were forgotten as he struggled for air and land.

Under any circumstances, falling into a flooded river in the winter would be a serious situation, but Kyle's predicament was greatly compounded by the chest waders he was wearing. Chest waders are bulky and cumbersome on land. In deep water they can be a death trap as they fill with water and drag you down. Fortunately, Kyle had his waders cinched rather tightly about his chest, so the flow of water into them was a trickle rather than a deluge. Kyle is a strong swimmer, but this wasn't a swimming situation. He was in a struggle to get to shore before his waders took on too much water, and the only technique was to flail and grope and scramble. It seemed to Kyle like he was in the water forever, but in less than a minute he had struggled to saplings growing on the riverbank and pulled himself out of the water.

Kyle's waders were filled with a couple of gallons of freezing water, which had now drained into the boots and legs. The air temperature was in the low forties, and he was already shivering, but there was no time for self-pity. As he stripped off the soaked waders and poured out the water, Kyle watched in horror as his kayak drifted down the Choctawhatchee. The paddle was gone—out of sight, never to be seen again. Then, by an incredible stroke of luck, his kayak caught in the branches of another fallen tree about 200 yards downstream. Kyle knew well that it wasn't just a boat hung on that snag—it was the rest of our field season and our best hope for documenting

Kyle paddling down the Roaring Cutoff with his feet up on the bow of the boat. (Photograph by Brian W. Rolek.)

ivorybills. In the dry bag that he prayed was still in the boat were more than half of the memory cards to run the listening stations, and in the bottom of the boat were more than half of our batteries. The kayak was borrowed from Mark. Kyle knew he had to save our equipment.

The boat was completely swamped so that only the tip of the stern bobbed above the surface of the water. The kayak was hung in branches only about 20 feet from shore, but it was in deep, fast-moving water. Kyle didn't hesitate. He stripped off his wet waders and the rest of his clothes, and plunged back into the cold swirling water of the river. A minute later, chilled to his core and still in a bit of a pickle, Kyle was lying on shore holding tight to the flooded boat as the current fought to pull it further downstream. It was only with great effort that he was able to roll the swamped boat on its side, draining most of the water, and haul it out of the river.

As he pulled the boat up on the bank, standing wet and naked in the empty forest, Kyle let out a cry of joy. The dry bag with the memory cards and all of the batteries were still in the boat. They hadn't fallen out. The field season wasn't ruined.

Kyle was shivering and could feel his movements becoming sluggish as his core temperature slowly dropped. But with the air temperature in the forties, he knew he wasn't in serious danger of hypothermia if he kept moving. He pulled on his dripping wet clothes, including his soaked waders. At first the wet clothes only made him colder, but soon the neoprene of the waders and the synthetic fibers of his shirt and coat started to trap his body heat. He wasn't exactly toasty, but he started to feel a little less frozen. He slowly made his way back to camp, paddling his kayak with his long arms.

All things considered, Kyle's boat mishap was only a minor disaster. One memory card and Mark's kayak paddle were lost to the river. We had an extra kayak paddle at camp and one extra memory card for the listening stations. Kyle had saved the kayak, other equipment, and, most important, himself.

The last two weeks in March were to be our big push to find a nest and capture an ivorybill on video. Tyler was to be on the site for nine days from March 18 to 26, and then Dan and I would be searching for three days, March 30 to April 1. After that, it would be mostly Brian and Kyle alone.

Tyler called on March 26 on his way home from his week of ivorybill searching.

"I guess you don't have a video or you would have called sooner," I surmised when I took Tyler's call.

"No video. No nest. Sorry. I was really hoping that I could come through for us," Tyler answered, clearly disappointed.

"I really thought you'd be able to find a nest or roost cavity for us," I agreed. Tyler has an uncanny knack for finding the tough bird under adverse conditions, and I had been counting on him once again working his magic.

"These birds are driving me nuts," Tyler continued. "Brian and I heard them almost every day I was in the swamp. And the weather stunk. It was windy every single day."

"Yeah, it was windy in Auburn all week, too. Where did you hear ivorybills? Did you hear both kents and double knocks?" I asked.

"More double knocks than kents, but I heard clear kents on two occasions," Tyler answered. "I mostly heard birds along Bruce Creek, including way up toward the landing."

"All the way to the plastic bag?" I responded, rather surprised. Someone had tied a plastic bag to a sapling in the creek at a tricky spot where much of the flow of the creek comes from the west but you have to veer to the north to get to Bruce Creek Landing. This spot is only about 250 yards from the landing and much farther up Bruce Creek than we normally searched.

"Almost. They seem to be hanging out in the swampy areas both north and south of the creek. I found a huge, newly cut cavity in the swamp to the south, and I watched it one night until dark."

"I don't guess you saw an ivorybill, or we'd be having a different conversation."

"No, but just before dark a Pileated Woodpecker flew in and went into the cavity. It was a huge cavity, too. I think having a pileated use this cavity might undermine our claim that cavities bigger than five inches in diameter were carved by ivorybills," Tyler added, clearly trying to encourage me to be cautious in how I presented what we had found so far.

"But we don't know that the pileated cut the cavity," I responded. "It just went into it."

"Yeah, I guess," Tyler answered, "but it looked very fresh. I don't know why an ivorybill would let a pileated use a new cavity."

"It does seem strange, but it's also hard to believe that pileateds cut cavities 50% bigger along the Choctawhatchee than anywhere else in their range," I said, pondering the confusing cavity situation.

"I'll have to let Brian figure out the cavities. I don't guess I'll be back to Bruce Creek for quite a while," Tyler concluded.

"Good luck with the end of spring semester and your summer job. We'll keep you updated on what we find," I said as I hung up the phone.

Tyler was now out of action—back to classes and then off to a summer job in the grasslands of Nebraska. It was up to Brian, Dan, Kyle, and me to try to get our picture.

At the end of March Dan flew to Columbus, Georgia with his father, Paul, and we drove down to Florida for a three-day visit to the study site. The water level of the river was way down on this weekend, meaning that it was hard to paddle but easy to walk around. I used most of this weekend to measure feeding trees. On the first morning, as I was hiking around pulling on bark, I followed an old roadbed west from Mennill's Hammock all the way out of the swamp. Just beyond a deep channel (now only 2 feet deep) that had stopped my progress in the past, I passed some signs indicating the edge of state land and entered a 15-year-old slash pine plantation. This pine plantation was immediately adjacent to a slough with tupelo and cypress and big cavities that indicated possible ivorybill activity.

I was interested to see if there were signs that ivorybills foraged in such a pine plantation. Two trees seemed like good candidates. Each was a recently dead slash pine, with brown needles still on the branches. Sections of bark on both trees had been scaled away in a manner consistent with ivorybill feeding. The bark on these pines was tightly adhered, as confirmed by my fish scale. A month later when I examined the planted pine stand behind the parking lot at Bruce Creek Landing, which is also adjacent to mature swamp forest with a lot of signs of ivorybills, I again found chiseled bark on pines that seemed to be the work of ivorybills.

The implications of ivorybills feeding in pine plantation were intriguing. Both of the pine stands in which I saw ivorybill feeding sign had been recently burned. Fire may kill or weaken some trees, allowing beetle infestation (although southern foresters often burn and thin their stands to prevent such beetle infestation), but just as important, I think, fire opens the pine forest, removing the dense midstory that develops in an unburned pine stand. With their large bodies and long wings, ivorybills likely prefer open woodlands, as James Tanner, Jerry Jackson, and others have suggested. From a conservation standpoint, use of slash pine plantation by ivorybills would mean that there is much more potential habitat to support Ivory-billed Woodpeckers on the Florida Panhandle than if birds were restricted to hardwood swamps. I'm certainly not suggesting that ivorybills venture far into pinelands away from swamp forest, but even if ivorybills only used commercial pine within 100 or 200 yards of the swamp forest, this would mean a lot of extra ivorybill habitat when multiplied across the entire Choctawhatchee River basin.

As the road rose in elevation, the slash pine gave way to sandhills covered in a scrub forest of sand pine and live oak. Walking from the hardwood swamp into this sandhills habitat was like walking from an oasis into a desert. None of the trees here was more than a few inches in diameter, and the vegetation was thick and impenetrable. The road through this scrub was drivable in a car, and it was obvious that local people used this area as a dump. There were more than two dozen piles of debris containing every imaginable form of household junk, from TVs, ranges, and a refrigerator to toys, wooden furniture, and kitchen trash. After spending so many

A 15-year-old slash pine plantation adjacent to a tupelo/cypress swamp where I saw evidence of ivorybill feeding. (Photograph by Geoffrey E. Hill.)

days in the pristine forest nearby, it was shocking to see the land used like a garbage can. I quickly retreated back into the swamp to measure more trees.

We failed to detect any ivorybills that weekend. On Saturday night, Dan, Paul, and I loaded my Honda and drove north to the Columbus airport to catch their plane home. It was now April, and although Brian and Kyle would be in the swamp for another month, we felt we had reached the end of our first year's ivorybill search.

"I have to be very pleased with what we were able to accomplish with a tiny search crew monitoring an enormous swamp forest," Dan commented as we cruised north on Highway 81, the sky to our left lit bright orange by the setting sun.

"Yeah, in most ways I'd agree that this winter and spring were a huge success. We relocated the birds that we found last summer, and we got loads of good sound recordings of them," I said. "And yet, I have this inescapable feeling that we failed. We did not find a nest, and we did not

Trash piled along a dirt road only a few hundred feet from pristine swamp forest. (Photograph by Geoffrey E. Hill.)

get a photo or video of an ivorybill. People are bound to raise the same questions that we keep asking ourselves: How could you be on site, living right next to the birds for four months, and never see an ivorybill perched and never get any sort of picture of one of the birds?"

"We definitely won't convince everyone with what we've got so far," Dan agreed. "But you know as well as anyone that there is a good reason we haven't gotten a picture of the bird yet."

"Yeah." I continued the thought. "It's a huge area and these are very wary birds with excellent hearing and vision that flee from people. Both Brian and I saw ivorybills at close range that had just come off a tree near us. The birds saw us and started to fly before we saw them. It is not easy to capture a fleeing bird on video. I think we need to come up with a better system for rapid photography next year."

"They certainly are much easier to hear and audio record than they are to see and videotape," Dan concluded.

"I imagine that people will compare our inability to get a picture of the bird with Cornell's failed effort to relocate the bird that Harrison and Gallagher spotted," I mused. "To that I can only reply that throughout the eleven months after we found ivorybills along Bruce Creek, we regularly detected two birds within a small area of forest. If I remember your last totals, we already have over 150 kents and double knocks recorded, and a lot more audio files to search. Seven of the twelve people who visited the swamp heard double knocks or kent calls. We also photographed and measured dozens of cavities with dimensions consistent with ivorybill cavities but too big to be attributed to pileateds. And we have documented numerous trees with scaled bark unlike any feeding trees that I've seen or measured outside of our ivorybill area."

"Yeah, we have a mountain of tangible evidence compared to what Cornell has gotten in Arkansas—at least compared to the evidence that they have released. The obvious difference is video. Without that, some people will remain skeptical."

"I'm perfectly content to call our evidence 'highly suggestive' rather than 'definitive,'" I concluded. "These birds aren't going anywhere. This was not a flyby, one-time sighting. If we are right about ivorybills living along the Choctawhatchee River, with a larger search crew, we'll locate roosts, nests, and get plenty of photos and videos of the birds. Just notifying the birding community is likely to result in pictures as more observers with cameras get into the swamp."

"I agree. The many old cavities suggest that ivorybills have been in this area for decades. This is a population that can be studied and managed," Dan said. "One thing that really nags at me, though, is how people were able to repeatedly find, approach, and watch ivorybill woodpeckers in the Singer Tract, when we seem incapable of getting close to ivorybills along the Choctawhatchee. Why are our birds so different?"

"I think the Singer Tract birds may have been quite an exceptional population of ivorybills," I began, reciting a theory I'd explained to my fellow ivorybill hunters many times. "There were some strange things about the Singer Tract birds. First it's amazing to me that the local residents in Louisiana paid so much attention to the ivorybills in that population. My experience is that very few people pay close attention to birds, especially in the south. Look at the people we meet around here. They have almost no knowledge of the birds in the swamp, even though they spend their whole lives fishing and hunting here. Kuhn, the local game warden who helped Allen and his crew and later Tanner in the Singer Tract, had monitored ivorybills closely for years. He already knew the location of nest trees and roost trees before any ornithologist from Cornell got to the site. I believe that the local people who lived near the Singer Tract may have habituated ivorybills to humans."

"Habituated? You don't think that collectors caused the rapid evolution of warier behavior in ivorybills?" Dan said with a smirk. We had talked about how unlikely it was for a rapid change in behavior among a population of birds to be genetically based. It made much more sense for the change to be cultural, like when white-tailed deer become very tame in just a generation when suburbs are laid out in their forests and all hunting stops.

"Yes, I do think the Singer Tract birds had become somewhat habituated to people," I continued. "An analogy I like is chimpanzees in Africa. Jane Goodall picked an accessible population of chimps in Tanzania that was already somewhat tame, and then she spent years getting these chimps completely habituated to people. We've all seen the pictures of her sitting among the chimps. From those films, one could get the idea that all chimps are tame, trusting animals that you just walk right up to and photograph while you're in Africa. If you visited most forests in Zaire or Gabon with Goodall's study in mind, however, you would be sorely disappointed. If you went to forests where chimps are thought to live—not national parks but forests where people hunt chimps—and went hiking around, you might hear some breaking branches or see some foliage shaking, but your chances of glimpsing a chimp would be slim. Your chances of taking an identifiable photo would be near zero. Chimps in these areas flee the first sight or sound or scent of humans. They are as wild and wary as any animal on earth, and they are very difficult to see. There would be abundant evidence of their presence (scat, sleeping mats, feeding areas), but the animals themselves would be extremely elusive. You can imagine coming back from a trip to Africa and telling your friends who only know wild chimps through the Goodall films that you are sure that you saw signs of chimps but couldn't get clear looks at the animals. They would likely respond that there is no way that you could be in the same forest as chimps and fail to get a look at them. They would say that if Jane Goodall was able to walk up and groom chimps, you should have been able to as well."

"You really think there are cultural differences among populations of ivorybills and that only the populations with wary cultures survived?" Dan asked.

"Yes. I think that it is entirely conceivable that different populations of ivorybills in different areas have different cultures regarding how wary they are of people. Most of the less wary birds were shot, except the Singer Tract birds. They had the trees cut from underneath them."

"Isn't that essentially the excuse that the Cornell group has used for why they can't relocate birds in Arkansas?" Dan asked.

"Yeah, it is, except it seems as if I've heard more about genetically-based change in wariness regarding the Arkansas birds—I'm not sure any biologist has made that speculation, though," I responded. "So far, we've been able to relocate birds consistently. If we can't find at least roost trees next year with a much larger search effort, I'd be the first to say that we must have stumbled onto wandering birds—or never had ivorybills at all. If a resident population of ivorybills is here, we'll find where they sleep and where they breed."

Cat Out of the Bag 13

A t the beginning of April 2006, we had to make a decision about how to proceed with our ivorybill study. I had hoped for a clear video of an ivorybill and with that the ability to announce definitive evidence for the persistence of ivorybills along the Choctawhatchee River in Florida. We fell short of that goal, which left us uncertain about our next move. Our search team had seen and heard ivorybills on our study site for a year. Brian, Tyler, or I saw birds in May, July, and December 2005 and in January, February, and April 2006. We had detected ivorybills on our site virtually every week since December. We found twenty cavities with entrance holes that were more than 5 inches high and dozens of feeding trees with bark more tightly adhering than any feeding trees that I could find away from the ivorybill area. And, most impressively, by early April we had about 150 sound recordings of kents and double knocks all from within our small study area. This number would rise to over 300 as Dan and his crew finished searching sound files through the summer, but even in April we had what we thought was an impressive body of sound evidence. Our evidence certainly seemed sufficient to warrant funding for a follow-up study. But what was our first step toward a funded follow-up and an announcement of what we had found?

I requested a meeting with Dean Schneller, and he suggested that we include Marie Wooten, a faculty colleague of mine in the Department of Biological Sciences at Auburn and associate dean for research for the College of Science and Mathematics. Marie was knowledgeable about the politics of endangered species work and knew the appropriate contacts at government agencies. I told Stewart and Marie that, although we had failed to get a picture of the bird, we had enough tangible evidence to apply for funds for more work in the area. I suggested that we approach the School of Forestry at Auburn for funding for another year, maybe even continuing to keep our discovery quiet until we could get a picture of an ivorybill. I also thought that perhaps we should contact the Florida Fish and Wildlife Conservation Commission and the Northwest Florida Water Management District to let them know that they had ivorybills on their land. But Marie advised another approach.

"The Forestry and Wildlife Department can't fund a project like this," she stated. "It will have to be funded through federal and state money. The Florida agencies will certainly have to be involved and will be key partners in any future work, but in my opinion you should start by notifying the [U.S.] Fish and Wildlife Service."

"You really think so?" I responded, a bit skeptical. "I'm a little leery of telling the Fish and Wildlife Service about our findings because of their close involvement with the Arkansas search."

"I have a friend, Chuck Hunter, in the Atlanta office who I'm sure would handle this right," Marie added.

"I know Chuck!" I blurted out. "You think he could be our primary contact?" Contacting an unfamiliar administrator at the Fish and Wildlife Service had always concerned me. From the start, I was worried that, in a blink of an eye, the Choctawhatchee River ivorybill study could be taken over by other research groups. We had already invested so much in this ivorybill project, and I was enjoying the swamp work.

Chuck Hunter was officially chief of refuges at the Atlanta office of the U.S. Fish and Wildlife Service (USFWS), but this title doesn't seem to capture what Chuck really does. From my perspective, Chuck seemed to be the USFWS point man on virtually all issues dealing with nongame bird management in the Southeast, especially as they related to habitat. Since my first few years at Auburn, I had been in occasional contact with Chuck regarding issues of southeastern bird conservation. As a matter of fact, only a month before this discussion with Marie, Chuck had written to get my opinion of Ted Kretschmann's sighting in Dadeville. Five years earlier I had been the lead author on the Partners in Flight Bird Conservation Plan for the Southern Cumberland Plateau/Ridge and Valley, and Chuck had come to Auburn for a couple of days to work with me to complete that project. Telling Chuck about our ivorybill discovery seemed like a good idea.

In early April I sent Chuck an e-mail saying that I had some information about ivorybills that required USFWS involvement. He called me on the phone within an hour.

"What have you got?" he asked, obviously curious about any ivorybill information that I considered good enough to warrant the e-mail I had sent.

I gave him the short version of how Brian, Tyler, and I had run into an ivorybill while kayaking. I told him that in follow-up visits, Tyler had gotten a clear view of the bird and that we spent the last four months trying to nail down evidence for their existence.

"What evidence have you amassed?" Chuck asked, clearly amazed that an ivorybill study had been going on completely under the radar screen of the dozens of university and government biologists who were meeting regularly about managing the ivorybill population in Arkansas as well as searching other areas.

"No video, unfortunately," I responded, feeling that I had to make that clear from the outset. "These are hard birds to get a picture of."

"Tell me about it," lamented Chuck. For the previous two years he had been assisting searches for ivorybills and habitat reconnaissance efforts across the Southeast, including the Cornell-led effort in Arkansas. The USFWS was as anxious for video evidence of an ivorybill as was any research team.

"But I think we have convincing evidence. There are many cavities and feeding trees in our area. Some of the cavities are the size reported for ivorybills by Tanner, and the bark scaling that we are seeing is like nothing I've seen pileateds do before," I continued, keeping my account rather vague.

"OK," Chuck responded, obviously looking for more than holes and missing bark. Everyone apparently thinks they've got ivorybill cavities and feeding sign. He was looking for better evidence.

"We've also got lots of audio recordings," I said, knowing this might get his attention a bit more.

"Lots?" Chuck responded.

"Yes. Over fifty recordings," I stated, knowing that this would get him to Auburn. We actually had more than seventy kent calls and more than seventy-five double knocks by this point in April. By the time we published our account, these numbers would climb to 210 kent calls and 99 double knocks. At this point, though, Dan was still compiling everything, and I thought that I should be conservative and say fifty.

"Did you say fifty!? Five-oh?" Chuck asked, now getting a bit excited. Apparently fifty was a lot. Chuck couldn't see it, but I had a huge smile on my face at his response. My little search group had been working in isolation for so long that we weren't sure how our evidence would be received.

"Yes. More than fifty," I continued. "A couple of good recordings from handheld cameras and microphones and well over fifty from listening stations."

"Listening stations? You mean Cornell's ARUs?" Chuck asked, thinking that only Cornell had the technology for remote sound monitoring.

"No. We built our own remote sound recording stations that we call listening stations," I stated matter-of-factly.

"What sounds have you recorded?" Chuck continued, letting the listening station issue go for the time being.

"Both kents and double knocks," I stated, knowing that this too would build our case for ivorybills.

To this point I hadn't told Chuck where our ivorybills were, and he hadn't asked. It seemed like he was assuming that I was talking about a location in Alabama.

"I'd like for you to come to Auburn so I can show you everything that we've got, and you can advise us on how to proceed. I would come to your office in Atlanta but my dean, Stew Schneller, has been involved in this with me, and he'd like to attend our initial meeting. Also, I think you know Marie Wooten."

"Certainly, I know Marie quite well."

"Marie is working with Stew and me, helping us figure out how to proceed with this woodpecker study. I'd like for her to be there, too. It will be Stew, Marie, you, and me if that's OK."

"That sounds fine," Chuck responded. "I will certainly come to Auburn. This is very exciting news." Then he asked the question that I'd been grappling with for the past few months: "Do you think your evidence is better than what's come out of Arkansas?"

I paused for a moment and then responded, "If we accept the bird in the Luneau video as an ivorybill, then definitely not. They have a picture of a bird and we don't. If we consider the Luneau video as inconclusive, then, yes, we have a much larger body of evidence. I'm anxious to show you what we've got."

"Believe me, I'm anxious to see it," Chuck said as we concluded our call.

We arranged to meet at 10 A.M. on April 20. Chuck was to come to my office, and together we'd go over to the conference room in the dean's new office complex. The day before, I sat down with Marie to plot strategy.

"How do you think we should proceed?" I asked.

"That depends on your objectives and concerns," Marie responded, clearly indicating that this was my deal, and she was just there to help.

"Our objectives are to get our evidence for ivorybills published without jeopardizing the birds and to get funding for a follow-up study," I stated. "My only real concern is that a rival research group will take over our study of ivorybills along this river."

"When you say 'rival research group,' you mean the Cornell group," Marie countered, not interested in hiding behind euphemisms.

"Yeah. It's a totally unfounded fear, really. I think that Fitz [John Fitzpatrick] and Cornell made a tactical blunder in claiming proof for the species too fast in Arkansas, but I have no reason to think of them as the enemy. Everyone I know who's worked with Fitz tells me what a great guy he is. I have absolutely no reason to believe that Fitz or anyone at the Lab of Ornithology is dishonest or so zealous that they would run over a research team like ours. Many of the people associated with the Cornell group are acquaintances and even friends of mine. Fitz has been focused on ivorybills for years now, almost single-handedly leading the resurgence in interest in the species. I'm pretty sure he will be ecstatic when he learns of our discovery in Florida and that he'll use our birds to rekindle excitement about the Lab of O search in Arkansas."

"So why make it your main concern?" Marie countered.

"Because," I continued, "the Lab of Ornithology has so many resources and so much invested in ivorybills. They seem to have run into a wall in Arkansas, and I feel that they can't help but be tempted to come over to Florida if they learn of a location where ivorybills can be reliably detected. They could claim that it is really my team intruding on their nationwide ivorybill hunt that they started years ago."

"Okay," Marie conceded, "so let's proceed cautiously."

"Should we tell Chuck where we're working?" I asked. This was the key point on which I was unsure.

"No. Not yet," Marie suggested. "Present your evidence and gauge his response. If he thinks it's convincing, then there will have to be a second meeting with his boss."

"Jon Andrew," I added. I had seen Jon listed as being in charge of the national ivorybill recovery team.

"Yes, I saw on the web that Jon is in charge of the recovery of the Ivory-billed Woodpecker. He will probably have more control than Chuck over how money is spent," Marie continued. "You can have a follow-up meeting with Chuck and Jon. At the meeting with Jon, let them know where you are working."

"Okay," I agreed. "I had intended to tell Chuck everything, but stretching this out a bit seems like a good idea."

"In the meantime get your paper ready for publication," Marie concluded.

"That won't be a problem. Dan and I already have a draft of a paper for *Nature*. I'm not sure they'll go for it, especially without a picture. If *Nature* doesn't want it, we'll have to restrategize."

"Whatever you want to do, but I thought you were telling me that rare bird reports should be assessed by bird identification experts," Marie said, repeating a gripe that I'd made to her.

"I know. I did," I admitted. "We're also planning to submit our findings to some people or organizations qualified to give an opinion on a rare bird report. I think I'll send the report to Jerry Jackson and Bill Pranty. Bill is chair of the National Rare Bird Committee. *Nature* won't recognize the reviews that we solicit ourselves, but we can post the assessments of the experts on our website so the public can see an honest discussion of the strength of our evidence."

"Okay, you work on your announcement as we start to get advice from Chuck and the Fish and Wildlife Service."

On the day of our meeting, Chuck called me at 8 A.M. to say that he'd made it through the morning rush in Atlanta much more easily than expected and was near my office—almost 2 hours ahead of schedule. I invited him up so we could talk before the meeting.

I hadn't seen Chuck in more than five years, but he had hardly changed. He had perhaps a bit more gray in his hair and beard, but he had lost none of his boundless enthusiasm for birds. Ivorybills were clearly a passion for him.

I could have been coy and withheld all of our ivorybill information until the formal meeting started, but I've never been good at following protocol or playing things close to the vest. Over the next hour I told Chuck pretty much all we had in the way of evidence. I only held back a few tidbits to allow a little drama in our meeting, and I didn't show him any photos or play any sounds since all of that was set up in the conference room.

Chuck seemed impressed when I told him that we were up to seventy-five double knocks and seventy kent calls from the listening stations and camcorders. He was also very interested in my bark-pulling method for quantifying feeding trees.

Chuck Hunter and Brian Rolek near the mouth of Bruce Creek a few weeks after we first told Chuck about ivorybills in the Choctawhatchee River basin. (Photograph by Geoffrey E. Hill.)

Just before 10 A.M. we moved over to the dean's conference room and were joined by Marie and Stewart. I started the meeting by summarizing our four lines of evidence for ivorybills—sight and sound detections, audio recordings, cavities, and feeding trees. Stewart had seen much of this before; Marie hadn't, but my slide show was really directed at Chuck. Only a true ivorybill enthusiast could get excited by pictures of holes in trees and the weird sound clips that we'd gathered. I played the series of nine double knocks that Tyler had recorded in December, telling Chuck I thought that this was our most dramatic recording.

"Has anyone else gotten a recording of double knocks this compelling?" I asked Chuck as the sound clip ended.

"No," he answered matter-of-factly. "That's better than anything I've heard before."

I then played a bunch of individual double knocks recorded by listening stations in January and February. Chuck was riveted.

"Listen to how slow they are," he commented. "*Campephilus* species in South America are faster. Are all your recordings following the loud-soft pattern typical of *Campephilus* woodpeckers?

"No. About half are loud-soft. In the other half both knocks are equally loud or the second knock is louder."

"Interesting," Chuck responded. "The Cornell guys will be fascinated by this. As you commented to me, many of their recordings have the second knock that is louder, and they don't know what to think of those."

I moved on to the kent calls and played the examples that Dan had set up. These were recorded by listening stations mostly in February. Again, Chuck listened to the sound clips intently.

"These are not exactly like the kent calls on the Allen recording, but they are close," he commented. "Of course, the Cornell researchers in the thirties said that the kent calls they recorded at the nest weren't really typical of what they heard out in the woods. Very interesting."

I then came to what I thought Chuck might find most intriguing of all. "On a few days we recorded multiple kents and double knocks from the same recorder or adjacent recorders," I stated as I flashed up the next slide.

"Wow," Chuck exhaled, as he stared at the slide.

Dan had summarized the multiple-detection days with time-line illustrations. Along the time line showing a 6 or 12-hour recording period, Dan indicated in red where double knocks occurred and in blue where we recorded kents. The first time line summarized detections from Listening Station 1 from January 12. On that day between 8 A.M. and 2 P.M. we recorded five double knocks and one kent call.

I then flashed to the next time line showing January 22, the day after I had seen an ivorybill and a day on which we recorded three kents and five double knocks, all at listening stations 2 and 4, which are situated 500 yards apart and within 500 yards of my sighting. Chuck stared intently at the slide, trying to take in all the information, but I didn't linger on this slide long. After a few seconds I flashed up one of our biggest recording days of all: February 6, during the week in which Brian and Kyle had heard so many kents. This slide was covered with red and blue detections. Between listening stations 2 and 7, situated within 500 meters of each other, we recorded six kents in three bouts and two double knocks between noon and 6 P.M.

Chuck seemed a bit overwhelmed. "Nobody has ever gotten anything

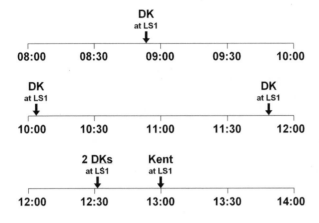

A time line showing when kents and double knocks were recorded by listening station 1 on January 12, 2006. (Powerpoint slide by Daniel J. Mennill.)

like this before," he stammered. "This is what Cornell had been hoping for in the White River."

"Really?" I asked, very pleased with Chuck's reaction. Of course, I knew that kents and double knocks together on the same sound recording was the most convincing sort of sound evidence.

"Absolutely," he continued, "never a kent and a double knock on the same recorder in the same day. Certainly not repeated kents and double knocks like you are showing."

I moved on to cavities showing pictures of some of our best with their dimensions and a frequency distribution of cavity entrance sizes. Chuck didn't seem as impressed by cavities, but I wanted to make a point that I thought that so many large cavities was quite intriguing.

"Didn't I read that they've found hardly any big cavities in the Cache and White river areas?" I asked, trying to gauge how unusual tens of big cavities in our small study area really was.

"They have found a few," Chuck replied, a bit defensive. "But, you're right. They have not found as many big cavities as you'd expect if there was a resident population of ivorybills."

"I read in that *North American Birds* article that they are using a 9-centimeter width as their criterion for considering a cavity suggestive of ivorybills. When we use a 10.2 centimeter width criterion, which is 4 inches, we have 66 cavities that exceed the cutoff within our little study area."

"So many large cavities is very interesting and very different from what they are seeing in Arkansas," Chuck admitted.

Then we came to feeding trees. I showed pictures of what I interpreted as ivorybill bark scaling. I explained that in our study area putative ivorybill bark scaling seemed to occur mostly on spruce pine and sweetgum, but that we found scaling on a variety of tree species. As I showed him pictures, I said, "These are the chisels most suggestive of ivorybills."

"What do you mean by chisels?" Chuck asked. I had forgotten that we had made up this use for the term just as we had made up our use for the term "hammering."

"By chisels, I mean places where very tightly adhering bark has been pried away from the wood. When ivorybills do this they seldom mark the wood. Their flat bills allow them to cleanly separate bark from wood," I explained.

"Do you see large pieces of bark under these feeding trees?" Chuck asked.

"No, not very large pieces," I responded. "This fresh, tightly adhering bark does not come off in big sheets. It has to be pried off one small section at a time."

"What you are showing me is different from the sort of feeding trees that other groups are focused on," Chuck commented. "I'm not sure what to make of it. I'm really fascinated that you mention pine at the edge of the swamps, though. Foraging on pine is mentioned so often in the old literature, it is very interesting that you are detecting these unusual feeding trees on pine."

"Look at this," I said as I brought up the slide showing where a sliver of wood on a feeding tree had been curled. "I don't think a pileated bill could curl wood like this. I think it could only be done by a bird with a flat bill."

"I don't know about that," Chuck responded. He seemed a bit overwhelmed by all the stuff I was throwing at him. "This is all so new. No one has presented feeding sign like this before—at least that I've seen. Very interesting."

As I completed the presentation, I asked Chuck, "So what do you think? Are you convinced?"

"The audio evidence is compelling," he said.

"What about everything taken together—the audio plus a dozen sightings, numerous giant cavities, and unique feeding sign?" I prodded.

"Yes, taken together it makes quite a story. I'm anxious to show this to Jon."

As we concluded our meeting, I had to run for my car. Long before the meeting with Chuck had been arranged, I had promised Wendy that I would visit her in South Carolina that weekend. I was now getting a very late start on my seven-hour drive to the Atlantic Coast. I thanked Chuck for coming, pointed my car east, and started my long trek.

Chuck called me a few days later saying that Jon Andrew was impressed by the evidence that we had amassed. They wanted another meeting so we could tell them where we were working, and they could help us contact state agencies, outline what funding might be available for a follow-up study, and advise us on the legalities of studying an endangered species. We arranged a meeting in Auburn the next week.

I was a bit more apprehensive about this meeting than I had been about the meeting with Chuck. For one thing, it was time for me to put my finger on a map and reveal our location. We had kept our location secret for so long, I knew it was going to feel strange to reveal our site. Also, I had never met Jon, and I knew he had been working rather closely with the Cornell team. I couldn't predict how sympathetic he'd be to keep my team in charge of studies of ivorybills along the Choctawhatchee River and to shield knowledge of our discovery from Cornell until we could make a public announcement.

Chuck and Jon got to my office about 45 minutes before we were to meet with Marie and Stewart. I found Jon to be an easygoing guy who conveyed a sense of honesty and integrity. He was someone I felt that I could trust, but I really had no choice at this point. The path forward was through USFWS, and Jon was the gatekeeper.

After introductions and a few minutes of chit-chat we sat down and got out maps. With no drama or build up, I told them where the ivorybills were.

"The Choctawhatchee," I said, just before I unrolled the maps. "Our ivorybill site is on the Choctawhatchee River. Is that what you would have guessed?" I asked Chuck.

"Actually it was my third of three choices," Chuck responded as he moved over to look at my map. "When you told me that you were in the Florida Panhandle and that you had spruce pine on your site, I knew that it was going to be the Chipola, Escambia, or Choctawhatchee. My first guess was going to be the Chipola."

"That would have been old news then, since John Dennis reported ivory-bills there in the fifties and sixties," I responded. "The Choctawhatchee is completely new. No one had ever mentioned the Choctawhatchee and ivorybill in the same sentence."

"Yeah, in retrospect, now that you've found ivorybills there, it is quite surprising that the Choctawhatchee River basin has gotten so little attention. I've often thought that those river systems in the Florida Panhandle needed more attention. We've been focusing elsewhere," Chuck said.

"Where exactly are you working?" Jon asked.

"Right here, where Bruce Creek flows into this cutoff, the Roaring Cutoff," I responded, putting my finger on the map at the spot where Beavertown had been.

"How much area have you been searching?"

"A very small area, given this vast forest. About two square miles. I can cover our entire study area with my fist," I said, as I literally did just that, pressing my fist over the study area on the map.

"How much have you searched surrounding areas along the river?" Jon continued, knowing that the difference between the discovery of a wandering ivorybill or two and a breeding population was huge.

"Very little," I responded. "We've focused almost all of our efforts on our little study site, about two square miles of forest. As I told Chuck last time, for most of the past four months, Brian Rolek was pretty much our entire search team. The other guy who lived in the swamp, Kyle Swiston, managed our listening stations, but he really didn't spend much time searching for ivorybills. I was there on a few weekends. We couldn't even manage to really cover our little study site. We had no time for off-site searches."

By the time we had looked at maps of the study area and discussed the scope of the search that we had conducted, it was time to move over to the dean's conference room and start our formal meeting.

At this meeting we first discussed the need for permits. Jon explained that we wouldn't need a permit until our evidence of birds was accepted. Even after the presence of ivorybills was established, we would not need a "take" permit until we started intrusive activities like monitoring a nest from a blind or climbing cavity trees. They suggested I start working on a permit that fall and winter.

Funding is what I really wanted to hear about. Jon explained that funding decisions for the next year, 2007, would be made in the fall by an advisory committee. He said that they would have a few hundred thousand dollars to fund all the ivorybill searches and other ivorybill activities around the country. He advised that once we went public with our Florida study, we could make a case for getting some of that funding for work along the Choctawhatchee River.

And finally we talked about how best to inform officials at agencies in the state of Florida that they had ivorybills on their land. Jon said that the next week he was scheduled to meet with Ken Haddad, executive director of the Florida Fish and Wildlife Conservation Commission. He said that he would let Ken know about our evidence for ivorybills along the Choctawhatchee. Jon and Chuck knew Ken well, and they trusted him to keep the ivorybill news quiet until we could publish our evidence. We agreed that we would not inform anyone at the Northwest Florida Water Management District until we got closer to the publication of our paper. Neither Jon nor Chuck knew anyone at that agency, and we knew that the more people we told, the more likely it was that word would get out before we were ready. The Northwest Florida Water Management District would have to be a key partner in future ivorybill work, but they didn't need to be brought on board just yet.

That was it. There was nothing else we could do until the publication of our paper got closer.

On the weekend after my meetings with Chuck and Jon, I paddled down to Beavertown for the final search weekend of the spring and to help Brian and Kyle break camp and come back to civilization. Brian had been in the swamp since mid-December and Kyle since early January, and they had had only a couple of days off in the entire four-month period. They were understandably ready to be back in civilization with clean beds, hot showers, and real toilets. They were also fantasizing about having something for dinner other than pasta.

The weather for this final weekend during the last few days of April was nearly perfect—mid-seventies, blue skies, low humidity. There were a few mosquitoes but not many. The spring drought continued, and the swamp was very dry—almost as dry as it had been in November 2005 when I walked around in tennis shoes. I paddled into camp early on Friday, picked up a video camera that Brian had ready for me, and then I hiked around in boots through the mature forest on the south side of Bruce Creek. I found a lot of big trees, including a few with cavities, but almost no feeding trees in this whole area. I detected no ivorybills. I looped back to camp and found Brian and Kyle taking a lunch break.

"Hey guys," I said as I looked for my lunch stuff. "Any luck the last two weeks?"

"We've had some double knocks and some hammering," Brian answered. "Oh yeah, and yesterday we saw what was probably an ivorybill fly across Bruce Creek."

"We? You mean you and Kyle both saw it?" I asked. "You know we haven't had any good multi-observer sightings yet."

"Yeah, I finally might have seen one," Kyle chimed in.

"So you guys were together in the canoe?" I continued.

"Yeah. We were paddling to Bruce Creek Landing to put a load of stuff in my car and this bird flew over the creek in front of us," Brian explained. "It flew straight and fast. It was a big woodpecker but it was not a pileated."

"Did you see a white trailing edge on the wings?" I asked, trying to get a feel for just how good this sighting was.

"No," Kyle responded, "the light was terrible. The sun was in our eyes, and all we could see was the shape of the bird. I don't know what it could have been except an ivorybill."

"Yeah, it was a tough look but the size and shape and behavior all said ivorybill," Brian added. "That part of Bruce Creek has been where we've been getting all of our detections since Tyler was here in March. I haven't seen or heard ivorybills in any of the areas south of us for six weeks. All activity seems to have shifted up around Bruce Creek."

"That's where I'm headed after lunch," I said. "I want to hike through the forest on the north side of the creek all the way to the landing."

"That's great," Brian responded. "We need to keep watching that area. You won't run into any water, that's for sure. This place is getting bone dry."

After lunch I paddled across Bruce Creek and then walked around in the woods on the north side of the creek. I found several hammocks with stands of spruce pine and sweetgum, and I started to find feeding trees where tightly adhering bark had been scaled from recently dead trees. I used my fishing scale to measure the bark on feeding trees—by this point I had recorded bark adhesion on almost 200 trees in our study area along the Choctawhatchee. I kept meandering east as I searched for trees to measure. Pretty soon I was close to Bruce Creek Landing. I ended up walking out of the swamp into a stand of Slash Pine that we can see behind the landing when we park. I found several recently dead slash pine from which tightly adhering bark had been chiseled cleanly away from the underlying wood. It looked to me like an ivorybill had fed in these pines. I spent the rest of the day hiking in the woods and measuring bark adhesion on feeding trees, but I didn't detect any ivorybills.

The three of us spent Saturday packing up camp and hauling everything up to the landing. It was sad to see Beavertown torn down. The camp had served us well. It had been home to Brian and Kyle for four months. Now we were giving it back to the ivorybills. I was anxious to see on subsequent trips whether there would be fresh ivorybill feeding sign on trees right in the old campsite.

For our last trip up the creek, we jammed a mountain of camp gear into the canoe. I was glad that Brian and Kyle had gotten skilled at canoeing because this overloaded boat was a disaster waiting to happen. We also packed the two extra kayaks full of gear. Kyle and Brian towed one of these loaded kayaks behind their canoe, and I towed the other behind my little green kayak. As I paddled the first hundred meters or so away from camp, I was relieved that it wasn't as challenging to tow the fully laden kayak as I feared it might be. Then I looked back and realized why it was so easy—the other kayak wasn't there! It had come undone just as I started paddling upstream. I retrieved the loose kayak a few hundred feet downstream and started over. It wasn't so easy the second time.

I was worried that we wouldn't have room in our two vehicles to haul all of our gear out of the swamp, but by piling things from floor to ceiling in both cars, we just squeezed it in. I had three kayaks on the roof of my CRV and Brian and Kyle strapped the canoe to the roof of Brian's Montero. Brian and Kyle trailed me north to my house in Auburn, spent the night, and then got an early start in the morning. Brian drove Kyle home to Windsor, Ontario and drove back down to his parent's house in Pennsylvania. He was going to take about ten days off and then come back to Auburn to work with me on the ivorybill project the rest of the summer. With Kyle and Brian gone and spring field work over, it was time to tell the birding and conservation world what we had found.

Return of the Lord God Bird   14

As I write this last chapter, we are preparing to publish our evidence that Ivory-billed Woodpeckers exist along the Choctawhatchee River. At present, we have strong evidence that from May 2005 to April 2006, at least two Ivory-billed Woodpeckers lived near the mouth of Bruce Creek.

Does this mean that the Ivory-billed Woodpecker persists as a total of two birds in the Choctawhatchee River basin? Almost certainly not. The Bruce Creek area where we monitored ivorybills in the winter and spring of 2006 is only one small area—about 2 square miles—of a huge floodplain forest. Our small search crew spent almost all of its time documenting ivorybills in the location where we first detected them. Through April 2006 we conducted almost no searches away from our main study area.

On May 19, 2006, Brian and I left Auburn driving my car and a university van to do a bit of reconnaissance above and below our study area. I had just submitted a summary of our evidence to the journal *Nature*, and I was ready to get out of the office and back into the swamp. Our goals were to assess habitat along a broad stretch of river and use our experience to find ivorybill cavities and feeding trees to look for evidence for additional birds. On Friday we put in at Highway 90 east of Ponce de Leon with the goal of floating south under I-10 and exploring the forest south of the expressway.

We planned to take out at Camel Landing, 7 or 8 miles below Highway 90. As we approached the Camel Landing area to drop off a vehicle, there was a small side road with a sign that said "Cow Lake Landing." *This must be it*, we thought, and drove a half mile through pine plantation to a landing that looked much like Bruce Creek Landing, with a ramp leading down to a still creek or oxbow, not the river.

"If we've learned anything this past year, it's that we can't assume that this oxbow connects to the river," I said as we got out of our vehicles. "We better confirm that we can float to the river from here before we leave a vehicle."

We got in our kayaks and floated in both directions away from the landing. There was no connection to the river. There were, however, several large cavities and potential ivorybill feeding signs on a dead sweetgum and a dead spruce pine along the oxbow. This spot is about five miles north of our Bruce Creek study site, and in my mind what we were seeing was evidence for an Ivory-billed Woodpecker that was not one of the Bruce Creek birds.

After our quick paddle around Cow Lake, we put our boats back in the van and went another half mile down the road, where we found a public boat ramp right at the river. We left my car and drove the van north to our put-in spot at the Highway 90 bridge. We got on the river around 9 A.M.

South of I-10 we left the main river and entered a small cutoff that took us deep into the forest. For the next seven hours we floated down this small channel past ancient oaks and spruce pine with an occasional stand of gigantic cypress. About a mile or so south of where we turned off the river, representing about three miles of paddling and a dozen portages around fallen trees, we found a stand of huge cypress with many very large cavities. These holes were huge, easily 5 inches in height in both of our opinions (although we had no means to measure them), and two of the holes appeared to have been dug that year because they had no scar tissue at all. There were probably 30 big cavities on five different trees in one little stand. Just beyond this stand of cypress, there was a small clearing next to the channel— the only break in this forest that we had seen—with a boat moored along the bank and a jeep parked in the grass next to a shed. It was strange to spend hours penetrating to the heart of a vast swamp wilderness only to end up looking at a parked car. This little private landing sat right on the edge of the swamp. There was more than a mile of mature swamp forest between us and the Choctawhatchee River at this point.

We continued down the channel, now a bit wider and with fallen trees cut to create a boat passage. About ten minutes after we passed the private landing, we heard a motorboat puttering toward us. We moved over to one

Brian exploring a remote section of bottomland forest along the Choctawhatchee River on May 19, 2006. (Photograph by Geoffrey E. Hill.)

side of the small channel as a tiny johnboat propelled by a little motor belching gray smoke came around a bend. In the boat were a middle-aged man and woman. The man was portly and bald and wore a weathered cotton shirt that was open to reveal a stained white undershirt. The woman wore a flowered cotton dress. I saw the man do an almost comical double take when he noticed us along the bank of the channel. He seemed determined not to show any surprise at our presence, although Brian and I later joked that we were probably the first outsiders he had ever run into in this channel. There was no clear boat access to this cutoff from the river at either end. We had spent more than 4 hours crashing through and pulling our boats around dozens of fallen trees blocking the channel. I imagine that these folks always thought that their private landing was the only access.

The man puttered over to a trotline and casually pulled an old bait off and put a new bait on the hook. Then he turned his attention to us. Brian and I were apprehensive. Brian told me a while later that it was too bad he had just watched the movie *Deliverance* for the first time a few days before. He said he couldn't get the rape scene out of his head—"Squeal! Squeal!" Our dealing with the local people for the past year had been almost all positive (except for Brian's stolen kayak), but we were now way back in the swamp in an area that these folks might think of as their private fishing hole.

"How's it going?" I said as the man finally looked at us.

He took a long pause as he eyed us. Then he spit a stream of tobacco into the channel and replied, "Must be a rough ride in dem things."

I let out a huge sigh. It was going to be fine. These folks were as friendly as everyone else along the river.

With a big smile I replied, "No, it's a smooth ride down the creek, except when we have to crash over logs."

Brian immediately followed with "How far is it to the river?"

The man took another long pause, let loose another stream of tobacco, and replied, "It's a fer piece. Where're you boys fixin t' git?"

"We need to get to the Camel Landing. That's where we left our car," Brian said.

Another pause, another stream of tobacco. "Ya ain't gonna make it."

I could see Brian's eyes get big. It was about 4 P.M. For the past five hours we'd been slowly moving along the channel searching for cavities and getting out and walking through the forest. We figured we could shoot down the rest of the channel to the river whenever we wanted and be back to the car in a couple of hours. We had until about 7:45 before it would be dark. It wasn't clear why this man was telling us we wouldn't get back to our car.

"What's it like ahead?" Brian stammered, wondering what challenge would keep us from reaching the river.

"Ya gots t' git through a bunch a trees before ya reach da Choct. An deys a bunch a chann'ls dat deadends or il git ya turnt around. It ain't easy t' find d'way."

"Geeze," I muttered. "Sounds like we better get moving. We have to get as far as we can before it gets dark."

"Did yas run inta any udder boats in t'crick?" the man asked as we drifted away.

"No, but there was a boat pulled up at the clearing back there with a jeep parked next to it," I answered.

"Oh, okay. That'd be Jimmy. We best git," the man concluded.

"Thanks for your help," Brian and I said as we started to paddle off.

The man nodded and spit another tobacco stream. We pulled away and shot through a tight spot where a tree had fallen over the channel. The man and his wife had just struggled to get their johnboat through this tangle.

As I turned back to wave good-bye, I could almost see a smile on the man's face as he watched us easily slip through the barrier. "Oh, yous'll make it. Dem things don't get hung."

We headed down the channel, glad to have his vote of confidence. I had never doubted we'd make it, but we decided we better get moving if we didn't want to end up groping for the car in the dark. Brian led as we moved quickly down the channel. For the next half mile or so trees that had fallen across the channel were sawed, and we only had to backtrack once or twice when we chose a channel that proved impassable.

About 45 minutes after we left the man and woman, I was following Brian as I paddled around a fallen tree when Brian suddenly pointed excitedly to his left exclaiming, "Whoa, Ivorybill!" He looked at me to make sure I had heard him and then turned back and raised his binoculars. I looked in the direction Brian was looking and waited but saw nothing. A couple of minutes later Brian said, "I think it flew off. Did you see it?"

"No," I responded. I had seen nothing.

We got out of our boats and walked into the forest for five minutes, but the ivorybill was long gone. We knew we had to keep paddling, so we climbed back into our boats and continued down the channel chatting about the sighting as we went.

"Geoff, that was one of the best looks I've ever had," Brian exclaimed. "It flew right across the channel in front of me. At first all I could see were white wings, but then as it got to my left into the woods I saw that it had a black head and back and front wings, with white on the back side of its wings. It swooped up onto the trunk of a tree and landed like a woodpecker. It was a huge black woodpecker with white secondaries."

I wanted to scream. I had missed another look at an ivorybill. I don't know why I didn't see the bird. I was probably looking down at the branches in the channel, navigating my kayak at the wrong instant. I'm not the most observant birder. I have a knack for looking the wrong way at the critical moment.

"You saw it land?" I asked, feeling like I'd had this conversation before with Brian and Tyler when they had seen ivorybills that I had missed.

"No, just as it landed I lost sight of it. There was a branch and some leaves in the way, but also I think it immediately hitched around to the backside of the tree. I looked at you to see if you saw the bird. When I looked back I thought I saw some movement in the trees. It was probably flying off."

"I can't believe I missed that bird," I mumbled mostly to myself. Then to Brian I said, "You know my video camera on the front of my boat was running."

"You think it was pointed the right way?" Brian asked hopefully.

"Let me see," I said as I rewound the tape and replayed the event on the little screen on the camera. This was a Sony Hi-8 camera, not a digital camera. The advantage of our Hi-8 cameras is that we had batteries that would last for 8 hours, and the tapes would record for 4 hours. The maximum battery time for the digital mini that Brian was using was 2 hours, and the mini tapes would only record 1 hour. I replayed the event, and the orientation of the camera looked good. Brian was framed in the center of the image as he led the way, and the camera view was centered over the channel when he raised his hand and pointed at the bird. On the little screen on the camera, however, I couldn't see any sign of a bird. I figured I should save the tape even though I couldn't see a bird on it. "Do you have another tape?" I asked Brian. Among my many bad habitats is to demand that my students and technicians be completely in charge of video supplies. I just expect to be able to put my hand out at any time and get a tape or batteries.

"No," Brian responded sheepishly, "there's one back in the van."

"Oh, well. I'll have to keep using this same tape. I'll just be careful not to tape over that ivorybill section." I could imagine half the birding world cringing at the thought of endangering a possible video of an ivorybill, but I wasn't going to put my camera away for the day.

We decided that I should be going first since we potentially would have had a video of an ivorybill if I had been in front of Brian when the bird flew past. I took the lead just in time for the scariest alligator encounter I've ever had. I frequently ramble about alligators being of no concern. I tell anyone who will listen that if there was a million-dollar prize for getting yourself bit by a wild alligator, it would be virtually impossible to claim in Alabama or the Florida Panhandle. Alligators are scarce, and they flee people as soon as they hear or see them. You can never get very close to them. "It is ridiculous to be afraid of alligators," I would state. But suddenly, fear of alligators didn't seem ridiculous at all. As I came around a bend in our little channel, a huge alligator slipped into the water. This animal was at least 10 feet long, with an enormous head and girth. The channel was barely 20 feet wide, and the current pushed me literally on top of this massive gator before I could do anything about it. As the tip of its huge tail slipped into the water, I was right over it, and it seemed like there wasn't room in the channel for both of us. I braced myself, expecting the alligator to thrash upward when the kayak struck its back, but nothing happened. It went down; the kayak went over. After that little incident, I decided to quit saying that alligators deserve no respect.

We finally made it out of the channel and back to the river around 6:15 P.M. The float down the Choctawhatchee River to Camel Landing took longer than we expected, but we got to watch three Swallow-tailed Kites swooping over the river and compare the tail lengths of the two adults to the shorter-tailed bird that appeared to be their offspring. We made it to the car just after sunset.

The next day we explored an area far to the south of our study site, in the delta of the Choctawhatchee River. We put in at Highway 20 just east of the town of Bruce. This location is about 15 miles downstream from our study site. We started by exploring an area called Reason Lakes just south of the highway, which is a series of oxbow lakes surrounded by tens of acres of huge tupelo and scattered enormous cypress growing out of standing water. Some of the large cypress had large cavities. The few hammocks in the area had large oaks and spruce pine, and on the one hammock that we searched we saw apparent ivorybill scaling on dead trees.

We continued southeast and explored Cowford Island. The part of this large island that we explored has some of the poorest forest I've seen in the whole river basin. This southeastern section of the island was high ground, maybe 10 feet above the high water marks on the banks of the river, and it was covered in puny oak, pine, and sweetgum. This forest seemed sad and pathetic after a year of searching majestic stands of oak and spruce pine. On some of the small pine trees, however, we found impressive sections of scaled bark. On about ten pines with diameters of 6 or 8 inches that had apparently been recently killed by bark beetles, large areas of tightly adhering bark had been scaled, revealing dozens of beetle bore holes. These were freshly dead trees with very strongly adhering bark. "If these aren't ivorybill feeds, then I've never seen an ivorybill feed," I declared to Brian. We took some photos of the feeding trees, walked around a bit, and moved on down the river.

After paddling along the main channel of the Choctawhatchee River, which is several hundred feet wide with little current in this stretch, we

One of the scaled pines that Brian and I encountered on Cowford Island with numerous insect bore holes revealed. (Photograph by Geoffrey E. Hill.)

turned down a channel called East River. On our maps the East River looked about equal in size to the Choctawhatchee River, but in life it was a small side channel with a swift current. It flowed through mature hardwood forest, and we found apparent ivorybill feeding trees and some large cavities, especially near a meander called Horseshoe Bend. At one point, as we were drifting down the East River just north of Horseshoe Bend, Brian pointed above the channel and said, "There goes a ducklike thing. Did you see it?" As usual, my response was "Where?" I hadn't seen a thing.

"You think it was an ivorybill?" I asked.

"I don't know," Brian replied. "It was big and flew like a duck."

"A wood duck?" I asked.

"No way. It was bigger and longer-winged," said Brian.

"Of the birds in this forest, only an ivorybill fits that description," I said. Then I added, "You know my camera is still running. It was pointed right at it."

"Wouldn't that be crazy if we got two ivorybill videos in 24 hours," Brian joked.

We ended the day paddling through what seemed like a giant lake where Pine Log Creek joins the East River. Just before we turned up Pine Log Creek, we came to an amazing abandoned logging bridge that spans the East River. The bed of this structure is only about 10 feet above the water of the channel and if it had been complete, it would have stopped all boats from plying the East River. We couldn't even get our kayaks under

it. It was abandoned decades earlier, and a rather small section along the
west shore was now cut away to allow recreational boats through. It stood
as a remarkable monument to the cypress-logging frenzy of the twentieth
century.

We turned up Pine Log Creek, which was quite wide at this lower section, and started our 2-mile upstream paddle to a landing. I joked with
Brian that Pine Log Creek was wider than the Ohio River. There was no
current, but a stiff breeze gave the lake some chop and made the last hour
of paddling a chore. We had left the big trees behind.
All the cypress and tupelo along Pine Log Creek grew
in dense stands of small trees.

As we paddled along, I commented to Brian, "You
know that ivorybill yesterday was your tenth sighting
in the last year. Your first sighting was May 21, 2005.
Today is May 20, 2006 so you just slipped that last one

*An abandoned logging bridge
over East River near Pine Log
Creek. This bridge leads to a
large, undeveloped island and
was constructed in the twentieth
century exclusively to remove
cypress logs. (Photograph by
Geoffrey E. Hill.)*

in under the wire." I knew exactly how many ivorybill sightings Brian had claimed because I had just compiled the list of sightings for our *Nature* paper.

"Ten sightings? I guess I never counted them up. Let's see, there was Hill's Swamp, and the first one and . . ." Brian started counting up his sightings in his head holding up fingers as he recalled them. "This is embarrassing but off the top of my head I can only remember eight."

"You've seen ivorybills so many times that you are starting to forget some of the sightings," I joked. "Yesterday's bird also had to be a different bird than the pair at Bruce Creek."

"Yeah, it was over 9 miles north. I agree that it had to be a different bird."

"That means that you've seen at least three different ivorybills," I said. "How many different ivorybills do you think we can now account for along the Choct?"

"Let's see. A pair at Bruce Creek. For argument sake, let's just assume a pair everywhere we've detected birds," Brian said, trying to get our detections organized in his head. "The bird yesterday would be a second pair, and we've seen good evidence down here."

"What about Cow Lake?" I asked.

"Yeah, okay, that seemed good too," Brian agreed.

"So four pairs?" I asked, trying to pin Brian down on what I thought was a low estimate. "As usual, my unofficial estimate is higher. First, I still think there is at least one pair on the west side of the river at Bruce Creek and one pair across the river along Carlisle Lakes. Then I'd say a pair at Cow Lake, two pairs in that vast forest along the channel we paddled yesterday, a pair at Lost Lake near Tilley Landing, a pair in the Reason's Lake area, a pair on Cowford Island, and a pair around Horseshoe Bend. That's nine pairs, and I'd say we've glimpsed less then 25% of this swamp. There could be tens of pairs in here. Of course, all of my estimates assume a home range size of about 4 square miles per pair. If each pair needs 10 or 15 square miles of forest, then my estimates would be way too high."

"I'm starting to feel more comfortable with your optimistic estimates," Brian admitted. "We found cavities and feeding trees so easily in these new areas, and there are so many areas covered in mature swamp forest. I can't believe I actually saw an ivorybill in a new area."

"I am extremely optimistic that the ivorybill can be saved. As a matter of fact, I think the ivorybill probably already saved itself in this river basin. All we humans have to do is figure out a way to count them."

We drove back to Auburn on Saturday night, and I asked Brian to try to get the videos converted to digital files so we could search them on our computers. It took Brian two days to find the right cables and get the video transfer to work. He called me late on Monday night with the news.

"Geoff, you videotaped an ivorybill!" he exclaimed.

"No way. You can see an ivorybill on the tape?" I couldn't believe my camera had captured an image of the bird that Brian had seen. I hadn't seen anything when I reviewed the tape on the camera.

"Yeah, the bird image is small, but I think it's pretty good. It definitely flashes a lot of white," Brian explained. Then he added, "You got that second bird too."

"The bird flying over the East River? We got two videos of ivorybills in one weekend?" I asked incredulously.

"Well, we got video of one ivorybill and one something flying over the East River. The East River bird's just a speck."

"Wow. Bring those clips to my office in the morning. I have to see them."

Brian was at my office the next morning at 7 A.M., just minutes after I arrived. He gave me a CD with two videos, and I cued them up.

The first clip was the bird flying over East River. It showed a tiny, indistinct image of a bird flying across the channel, but on several frames there was a conspicuous patch of white on the upper wings. At the points where the video caught the bird with its wings spread it seemed to be a big, long-winged bird with a long head and neck and tail.

"I think this flyover video is intriguing," I commented as we watched the clip for a third time. "I think this actually might be an ivorybill, but I don't think any skeptic will be impressed by this. We better focus on our other video."

We cued up the video of Brian's sighting. I watched as Brian paddled along the channel. The camera mounted on the front of my boat captured a smooth image of everything in front of me. On the tape you could hear my voice rambling on about feeding trees as we moved along. "I'm going to have to get rid of the audio on this," I thought. Just as Brian paddled under a tree arching over the channel, a bird flashing white zipped to the left. It is only visible for about 12 or 15 frames, but it is a large bird with a lot of white on its upper wing. On two frames it looks like it has a black leading edge and white trailing edge to the upper wing. It is a tiny, indistinct image, and I could not be certain of the position of the white. The underwing seemed

One frame from the video that records Brian's sighting of an ivorybill. Circled is the bird that flew from above Brian's head and that appears to have a white trailing edge on black wings. The image is too indistinct to be positively identified. (Photograph by Geoffrey E. Hill.)

to be completely white and the ventral plumage appeared black. What made the video compelling was that in life Brian had seen diagnostic plumage features of an Ivorybill-billed Woodpecker.

We spent the morning fooling with the video. We dumped it into my Macintosh computer using iMovie software and the images appeared slightly clearer. From iMovie we could capture frames, but we couldn't zoom in on the video. I called around campus and found a video lab that said they could help us. We took the two tapes over, and they proceeded to zoom in on the birds. In some ways zooming helped, you could maybe see the distribution of white on the wings more clearly, but mostly it just created a bigger blurry blob. At one point as the video technicians were working to capture a zoomed image, the tape was playing at normal speed and orientation on a big screen TV. As Brian and I watched the video play on the big screen, our target bird went to the left and then something else moved across the screen.

"What was that?" Brian and I both exclaimed.

The video tech backed up the clip and replayed it at half speed. We watched the bird flashing white move to the left above Brian's head, and then we watched in amazement as a second bird flew from the upper right toward the lower left portion of the screen as it crossed the channel and then dramatically swooped sharply upward as it entered the trees, flashing white on its back when it swooped.

"There were two ivorybills!" I exclaimed as we played the tape yet again. "You saw the second bird. Watch your head. You track the flight of that bird and then point at it. You didn't see that first bird."

"You're right," Brian agreed, as he stared at the images. "I always thought the orientation of that first bird in the video was a little funny because I watched an ivorybill fly across the channel. Now it makes sense. I floated right under that first bird and it flew off without me seeing it. It was the second bird I spotted and watched fly across the channel."

"Wow," I said, "this must have been a pair or a parent and young. The birds on these tapes are tiny and indistinct, but you positively identified that second bird as an ivorybill in life. With audio evidence of two birds on Bruce Creek, we now have evidence for at least four ivorybills. I just wish this video was clearer."

"Yeah, maybe we shouldn't be running the cameras set for long recording," Brian commented rather off-handedly

"What!" I exclaimed. "The camera wasn't set for the best quality picture?"

"No. I set it so it would record longer on a single tape, so you could tape 4 hours without rewinding," Brian explained.

"I didn't know that. I was rewinding my tape about every 2 hours. I thought we were at maximum resolution," I responded, frustrated that I hadn't discussed this with Brian before the trip.

"Sorry, but I thought you wanted to tape all day."

"No wonder this thing is blurry. We were using used tapes too." With a minimum budget for this project, we had no funds for new video tapes. All season we had been recycling old tapes used years earlier to film House Finch mate choice trials.

"Not just used tapes. All of those tapes got wet during our last few days in camp. I just got them dried out. Maybe we should stop using those."

"I taped these ivorybills on a used, dried-out tape with the camera set at minimum resolution? No wonder the image is so indistinct." I couldn't be too mad because I had been too lazy to oversee my own video equipment. Still, it was frustrating to think that we might have a much sharper image of an ivorybill with better video camera operation.

"You know," I said as much to myself as to Brian, "I thought that if one of our video cameras ever captured an image of an ivorybill in focus, we'd be done. We'd have our proof. But here it is, a bird in focus in the video image, and I'm not sure we've got it. I don't think the bird's image is clear enough to constitute proof."

After we found the second bird in the clip, Brian searched a much longer video segment from our float. He started finding birds flashing white in their wings going back a full 4 minutes before the spot on the tape where he points at the ivorybill. Most of the additional images are fleeting glimpses, but at one point a large bird with extensive white on its back swoops up onto the trunk of a tree. This bird moves 10 feet or more up the trunk before it flies off, showing a dark belly and extensive white on both the upper and under wings. Again, the image of the bird is in focus in the frame, but it is a small and indistinct image. Still the bird's behavior and the amount of white it shows are intriguing.

Brian also found evidence for a third bird in the few seconds of video that culminate with him positively identifying an ivorybill. Just before the bird comes off the tree above his head, there is a flash of white going left to a tree and then right out of the picture. You can't see any detail, but it looks pretty much like all the other birds flashing white that we are calling ivory-bills. This indistinct bird goes left, then the bird flushes above Brian's head, then the bird that Brian identified flies left. If we count this first image as an ivorybill, then there were at least three ivorybills along this stretch of creek. Three ivorybills together strongly suggests breeding. Late May is just when you'd expect there to be family groups of ivorybills if this was a breeding population.

Over the next week, I must have replayed the video clips 100 times. I was very uncertain about what to do. It was quite ironic that after a year of searching, we had ended up in essentially the same predicament as the Laboratory of Ornithology team after their first year. We had our elusive video, but it was far from a clear image of an ivorybill. And yet, on some frames it seemed pretty definitive—a white trailing edge on the upper wing, extensive white on the underwing, a dark belly. I could close my eyes and see myself in front of an audience stating as I pointed to the best frame, "This can only be an Ivory-billed Woodpecker." But I knew that wasn't so. It was consistent with an ivorybill, even quite suggestive of ivorybill, but this lousy video was far from conclusive proof. I had to rate this video as a little better than our feeding tree and cavity evidence and not nearly as good as our audio recordings. The most compelling aspect of the video was that Brian had seen one of the birds clearly and had positively identified it as an ivorybill.

The last thing the birding world needs right now is another bad ivorybill video, I thought. *Maybe we should just sit on this and go with our sounds, cavities, and feeding trees.*

But if we accept this video for what it is—suggestive of an Ivory-billed Woodpecker—and don't overinterpret it, I decided that we could present it as further evidence for ivorybills along the Choct as well as evidence that we were dealing with at least five different Ivory-billed Woodpeckers in this river basin.

I end this book in the middle of the adventure. As we enter the second year of our ivorybill study, virtually everything remains uncertain. I don't know where or how we will make a public announcement that we have found ivorybills. I don't know how the birding community, the academic community, or the community of wildlife managers and state and federal administrators will respond to our announcement. I don't know what sort of funding we'll have for a follow-up study or if an Auburn research team will remain the primary researchers monitoring this Choctawhatchee population of ivorybills. I can only guess at the number of birders who will want to come to this area to try to see these woodpeckers. The human side of the story could not be murkier.

What I am sure of is that the ivorybills are there. Not one bird. Not a single pair. At least a half dozen pairs and perhaps tens of pairs of Ivory-billed Woodpeckers in the extensive swamp forests along the Choctawhatchee River. The Ivory-billed Woodpecker is not extinct. It isn't even hanging by a thread. It has a solid toehold in the forests on the Florida Panhandle. My sincerest hope is that as a society we now do what we must to make sure this population expands to fill at least a portion of its former range. In an age of strip malls and 7-minute traffic lights, iPods and Islamic Jihads, more than ever we need deep, impenetrable swamps presided over by the Lord God of birds.

EPILOGUE

How to Be an Ivorybill Hunter

The discovery of a population of ivorybills along the Choctawhatchee River in Florida is an exciting and fun birding event. I think everyone who likes to travel and see new birds should have a chance to search for ivorybills in this beautiful swamp forest. When I was much younger, I used to think that rare animals should be totally shielded from humans. "Make giant parks with no human access and let the wildlife prosper," I would proclaim. "People will only degrade the wonder of a natural area." But as I've matured, I've come to understand that wildlife and areas of natural beauty are pointless unless humans can appreciate them. Only humans have an aesthetic sense. Without people the world would not be beautiful or wonderful; it would just be a big chemical reaction running with no purpose.

I started this book by saying that my ivorybill story had about ten main characters. Through the summer of 2006 that was a fair characterization. But the cast of characters involved in the Choctawhatchee River ivorybill search will expand tremendously when we tell the world about these birds. Now you have a chance to insert yourself into the story.

It's inevitable that when we make a public announcement about our discovery of ivorybills, a number of birders will want to rush to the Choctawhatchee River to try to get Ivory-billed Woodpecker on their life lists. I can't help but picture in my mind a herd of sheep running toward a feeding trough. Let me suggest that you resist the temptation to be a sheep. The ivorybills of the Choctawhatchee River bottomlands are not the same as a La Sagra's Flycatcher in a backyard in Miami that is likely to stay one or two days and then be gone from the continent. Evidence suggests that what we have discovered is a resident population of ivorybills. The many old cavities on our study site suggest that ivorybills have lived here for decades, and we can expect that they will be resident along the Choctawhatchee River for decades to come. Running down to Ponce de Leon like a little kid chasing an ice cream truck the day after we announce that we've found ivorybills will not increase your chances of seeing one of these magnificent creatures; it will not help your standing in the birding community; and it might hurt our efforts to preserve these birds. A successful trip to see ivorybills along the Choctawhatchee River will take planning and preparation, and the timing will be crucial.

The announcement of ivorybills in Arkansas and the rush to the Big Woods by many birders may have taken the edge off the response that we will get to our announcement of ivorybills in Florida. By now, everyone is aware that searching for ivorybills in a swamp forest is not like waiting in someone's backyard for a rare hummingbird to come to a feeder. The Ivory-billed Woodpecker is a difficult bird to add to your life list. Your chances of seeing an ivorybill in the Choctawhatchee River bottomlands are relatively small. I spent about thirty full days in the study area between December 2005 and April 2006, totaling around 300 hours of daylight, and I had identifiable looks at ivorybills twice. Tyler's and Brian's rate of visual detection is roughly the same—one look per 150 hours (fifteen full days) of searching. Our rate of detection of ivorybill sounds is higher. I estimate that I have heard a double knock or kent call about once per 50 hours (five full days) on the site from January to April. These rates of detections of ivorybills are pretty low compared to the rate at which I have detected most North American birds, but they are orders of magnitude higher than the rates of detection reported from the Big Woods in Arkansas. For instance, Martjan Lamertink, who is an expert at finding and observing large woodpeckers, spent many hundreds of hours deep in swamp forests of the White and Cache rivers in 2005 and 2006 in what are thought to be the best potential areas for ivorybills, and he has yet to glimpse or hear an ivorybill in Arkansas. An observant birder visiting the Choctawhatchee River for a full week of diligent and quiet searching in the winter or spring has a reasonable chance of hearing, if not actually seeing, an ivorybill.

We really want birders and nature enthusiasts to play the role of ivory-bill hunters rather than the role of sheep. Unlike the Cache and White rivers in Arkansas, the federal government owns no portions of the Choctawhatchee River basin. The State of Florida controls all of the public lands in the basin, and its priority to this point has been to protect the watershed to ensure clean water flowing into Choctawhatchee Bay and to provide for the recreational needs of the local population. The state's stewardship of this land over recent decades was directly responsible for the preservation of the population of ivorybills that we discovered. State agencies are not anxious to close off areas to hunting and fishing, which would incite negative attitudes toward Ivory-billed Woodpeckers and toward birdwatchers among local sportsman. Good public relationships with residents of the Choctawhatchee River basin will be key to the long-term conservation of ivorybills. Deer hunters and fisherman do nothing to endanger these woodpeckers. As I write these words, the tentative plan by the Florida Fish and Wildlife Conservation Commission is not to restrict access to any portions of state lands along the Choctawhatchee.

For obvious reasons, we want to avoid having lots of birders and other woodpecker chasers crowding into one or a few small areas. If many birders started crowding into our Bruce Creek study site, that would obviously drive ivorybills out and destroy the area. Instead of the heavy-handed and expensive approach of closing the area and patrolling with law enforcement agents, we want to limit access simply by appealing to birders to police themselves. My appeal to birders is simply to not enter the Bruce Creek area during the winter and spring of 2007. There is a vast swamp forest in the Choctawhatchee River basin to search for ivorybills. There is no need to crowd around Bruce Creek.

We don't want to avoid crowds just in the Bruce Creek area; we want to avoid concentrating birders anywhere. Spreading birders out in the swamp will minimize disturbance to any pair of ivorybills and will help us gather information on the distribution of these birds in the river basin. We ask the birders who hear or see ivorybills not to publicly announce the exact locations of their detections. Such an announcement is likely to cause a follow-the-leader response by birders, and we'll have too many people crowding into a small area and impacting birds. For the good of the ivorybill, we can never make this a stakeout bird and an easy tick on the checklist. Each individual and small group of birders is going to have to seek these birds in the swamp wilderness and find them for themselves. We ask ivorybill searchers who want to announce their success to post the locations where they see or hear ivorybills only in the most general terms and to send details of the detection to the research and recovery team.

We hope to use birders as a volunteer army to help us census ivorybills along the Choctawhatchee River and its tributaries. Searchers can send us reports if they see or hear ivorybills. For this to be of value, we ask all searchers to note the coordinates of their discoveries using GPS units. If you leave the main river channel when water is high, you will need a GPS unit (or a compass if you are good at traditional navigation) for your own safety. As an example of the value of noting locations on a GPS rather than relying on verbal descriptions, I found a cavity within a half mile of our Beavertown camp in November 2005 before I was using a GPS, and we never relocated this cavity despite weeks of continuous work in the area. The tree may have fallen down, but more likely we just never found this cavity again. Descriptions of detections will be of limited use without GPS coordinates. With a GPS, whenever you hear double knocks or kent calls or observe an ivorybill, you can send us the GPS coordinates, and we can enter them into our database and conduct follow-up searches of the area. If you get an audio recording or video image of an ivorybill, we would certainly like to see it or hear it and know where it came from. In this way we can harness the skills and energy of the birding community and rapidly develop a foundation of information from which we can assess how widespread ivorybills are in this river system and how many birds constitute the population. We will not just "allow" you to look for ivorybills but actually make you part of the effort to conserve these birds.

I think that the birding community is likely to stay interested in our study of the Choctawhatchee River ivorybills, so we will post regular updates on the ivorybill study. I know from the complaints that I heard about the Arkansas search that ivorybill enthusiasts want to be informed and that a public perception of secrecy regarding the Arkansas search alienated some birders. We are committed to sharing everything that we discover throughout the study. During the main field season, we will post weekly updates on our website stating what has been found or heard or seen and the general location where events happened. If, by the end of the first year, we start to have enough information to know areas where ivorybills are scarce and areas where ivorybills seem to be regularly present, we will publish distribution maps to help would-be ivorybill hunters. These will never be specific locations. Rather, we'll publish drainage-wide maps with blocks of forest shaded as "high use area," "moderate use area," or "low use area." Our pact with birders will be that birders behave responsibly (and out of the hundreds of birders I've met in my life, I've only met a few who would put getting a new bird on their list above the safety of birds), and we'll keep no secrets except specific locations of bird encounters.

I'll reiterate what I mentioned earlier in the book: imitations of ivorybill calls and knocks should absolutely never be used in the Choctawhatchee River basin. Members of my search team have not used them. There is no evidence that such sounds have any positive effect on your chances of seeing an ivorybill, but such sounds will corrupt our monitoring efforts and will mislead other birders into thinking they have detected an ivorybill. Attempting to imitate double knocks on the side of a boat or by banging on a tree would, in my mind, put you in the category of a birder who goes after a bird at all costs. I think we can all agree that when we hear a kent call or a double knock in the forests along the Choctawhatchee River, we want to be confident that it is an ivorybill and not a human imitating an ivorybill. Let the ivorybills do the banging. You do the listening.

Ivorybill hunting can be pursued at any level of difficulty, from a morning spent sitting and listening at a landing, to a drift down Holmes Creek or the Choctawhatchee River, to a multiday excursion deep into the swamp forest. The chances of detecting an ivorybill will tend to increase the deeper you get into the swamp, but as long as you are near swamp forest in this river basin, you have a chance of detecting an ivorybill. Be careful if you venture off the roads and away from landings. The Choctawhatchee River bottomlands form a vast swamp wilderness. Only when water levels drop to their lowest annual points with the river and creeks confined to their banks is the forest relatively easy to walk through, and only then does it become relatively hard to get lost. For most of the year, the entire basin is a maze of channels and dry spots, and it is very easy to get lost either on foot or in a boat. It is possible to get lost even navigating with a GPS. If you are in waders, you can find your path back to your starting point blocked by deep water, and if you are in a boat, you can find your path blocked by land or shallow water. Or you can drop your GPS into the swamp, which Murphy's law dictates will happen exactly when you reach the deepest portion of the swamp or are most confused about your location. The rule of thumb in this swamp is caution. There are plenty of snakes and alligators, but your real concerns should be drowning or getting lost.

The best way to see or hear an ivorybill will be to use a green or brown

kayak with all dark paddles (nothing is more conspicuous in a swamp forest than flashing white or silver paddles). If you've got a kayak that is powder blue or lemon yellow, now is as good a time as any to camoflauge it with
a can of green or brown spray paint. A kayak is the best way to search for
ivorybills. Both of my encounters were from the seat of a kayak. Alterna-
tively, canoeing along creeks, bays, and sloughs is a quiet way to penetrate
into ivorybill habitat, but a canoe will not get you to as many spots as a kayak
will. Use a GPS to track your route so you can find your way back after you
meander into the forest. I would recommend that you do not try to learn
how to maneuver a canoe or kayak in the swamp. Kayaking or canoeing re-
quires some upper body strength and conditioning, and it takes practice to
be able to maneuver either craft. Get out on any local body of water several
times at least before you try paddling in the ivorybill site. Also, be honest
about your level of physical fitness and plan your trip accordingly.

Walking can be as effective a strategy as paddling if the water level of
the Choctawhatchee River drops below about 6 feet at the Caryville gauge.
You can check the level of the river on a website run by the federal government (http://ahps.srh.noaa.gov/ahps2/hydrograph.php?wfo=tae&gage=carf1
&view=1,1,1,1,1,1). Between water levels of about 5 and 8 feet you can walk
a long way in some parts of the swamp, but there will be water channels
more than chest deep blocking many routes. When the water is higher than
8 feet, you should have a boat or stay at a landing. During cool and cold
weather, you will probably want to walk in chest waders so small pools of
water will not stop you. Don't attempt to cross deep creeks or flowing channels in waders. It is easy to have your feet swept out from under you. Also
be aware that walking in this swamp in waders is tricky. You are not walking across a flooded grassy lawn. The floor of a flooded forest is strewn with
sticks, logs, and vines. Rarely can you see your feet in water more than a few
inches deep, so it is a constant struggle not to be tripped. There are also
numerous holes and steep, narrow channels. Walking in waders through a
flooded swamp forest is slow, tedious business. When it warms up, you can
just hike in jeans and get wet when needed. Much of the forest is easy to
traverse on foot when the water is low.

Looking for ivorybills flying or perched is important, but mostly you
will want to listen. The most common noise we've heard ivorybills making
is hammering—very loud, rather slow bangs on trees. An ivorybill that is
hammering sounds like no bird you've heard before. You will swear someone is cutting wood with an axe or just pounding on trees with a bat. If you

detect such hammering, get your camera or binoculars ready and approach slowly. Ivorybills are extremely wary, and the chances are high that you won't get very close before the bird flushes away from you. Some birders are good stalkers, and if you are the sort of person willing to creep or belly crawl toward a foraging bird, you might be rewarded with a good look at an ivorybill. Ivorybills seem to have exceptional hearing as well as sharp eyes, so they are not easy to sneak up on. Brian, Tyler, and I have approached several hammering ivorybills, and the best we've done is to catch a glimpse of a bird as it flew away from us.

Of course, you'll mostly want to listen for kent calls and double knocks. North of Mexico, only Ivory-billed Woodpeckers give a true double knock. The double knocks of ivorybills sound like a large bird thumping a big tree. They are usually, but not always, "BAM-bam" (loud-soft), given rapidly. The two knocks come about as fast as a human can say the words "one, two." Other woodpeckers, especially Red-bellied Woodpeckers and Yellow-bellied Sapsuckers, occasionally give two taps that a hopeful birder could interpret as a double knock, but they never give a perfect ivorybill double knock. The double taps of smaller woodpeckers always lack the heavy mallet-on-beam quality of an ivorybill. Listen to the examples of double knocks on Dan Mennill's website (http://web2.uwindsor.ca/courses/biology/dmennill/IBWO/) before you go on an ivorybill search, and be conservative in what you call a double knock. If you hear double knocks, try to record them with a video camera.

Along the Choctawhatchee River, Ivory-billed Woodpeckers seem to give kent calls mostly from February to May. The kent calls given by ivorybills along the Choctawhatchee River do not sound exactly like the kents on the old Singer Tract recording. Listen to the examples of kent calls on Dan Mennill's web page as well as the kents recorded in the Singer Tract. There are no White-breasted Nuthatches along the Choctawhatchee River, and we detected no Blue Jays deep in the flooded forest during the winter. Because we've only experienced one winter in this swamp, we don't know if scarcity of Blue Jays in the swamp during the winter of 2005–2006 was a consequence of high water or if Blue Jays always vacate this swamp in winter. Don't assume that Blue Jays will be absent, and be wary of kent-like sounds being given by Blue Jays. Kents generally come in series, but unlike the rapid string of kents given at a nest in the Allen recordings in the Singer Tract, most of the kent calls that we've heard have been separated by several or tens of seconds. Several kent calls is what you need to hear if you want to be certain that you are not hearing a gray squirrel or a kent-like sound by another animal.

The best time for an ivorybill search is from December to May. The weather is usually pleasant during those months on the Florida Panhandle, although it can range from bitter cold in January to blazing hot in May. These months are likely the primary period for courtship and nesting for ivorybills, and ivorybills tend to be noisier and more conspicuous during these months than any other time of the year. Trees begin to leaf out in late February in this part of Florida, and there is dense foliage by the end of March, so cavities are easiest to find from December to February. None of this information is original. James Tanner gave exactly the same advice to ivorybill hunters in his book published in 1942. December can be an excellent month to search—we got our best recording of double knocks in late December. June can also be a good month, but you can count on it being hot, and mosquitoes can be a problem. From April to June there should be recently fledged ivorybills present, and such naïve birds should be the easiest birds to approach. Other months of the year are tougher for ivorybill hunts, although the birds are still in the forest, they still forage by banging on trees, and they still fly to and from their roosts. Brian saw two ivorybills at the end of July.

I'm very excited about the prospect of having some of the most skilled and enthusiastic birders from around the country searching the swamp forests along the Choctawhatchee River and Holmes Creek and sending us the evidence for ivorybills that they find. With a full year of effort, we failed to capture a clear picture of an ivorybill. Brian and I were the searchers with the most chances to film an ivorybill, and I will be the first to admit that we are both poor videographers. I'm confident that good pictures of ivorybills will be obtained once competent nature photographers get into this area.

We spent a year gathering a large amount of evidence that at least five Ivory-billed Woodpeckers persist in the forests along one small section of the Choctawhatchee River in Florida. It is now critical for those who want to protect and recover this magnificent bird to know how widespread ivorybills are in the Choctawhatchee River basin and throughout the Florida Panhandle. How do we best search new areas for the presence of ivorybills? I think that I've learned enough in the year that we've spent documenting ivorybills in the Bruce Creek area of the Choctawhatchee River to suggest some guidelines for how to conduct the most efficient searches when faced with enormous tracts of hard-to-access forest.

Ivorybills along the Choctawhatchee River, and I would presume in all of the areas they currently occupy, are very shy, quiet, reclusive birds. For whatever reason, they do not want people to get close to them. Their sharp eyes and sensitive ears allow them to detect searchers at a substantial distance and to move away from them. Most of the time, they make no sounds. For these reasons, trying to census for ivorybills by actually detecting the birds will be inefficient and ineffective.

Listening stations or autonomous recording units are also an ineffective means to search new areas for ivorybills. Listening stations are expensive. For the price of one of Cornell's audio recording units, a search team could hire a human searcher for several months. Moreover, listening stations will consume 1 hour of technician time for about every 6 hours of field recordings, and of course the sound technicians have to be trained and supervised. These investments become worthwhile when a search team has located an area with good preliminary evidence for ivorybills. In my opinion, however, using listening stations as a means to conduct primary searches of forest tracts is a poor use of resources.

The initial search of an area is best done by humans who know birds well and who are familiar with all animal sounds in southern forests. A search crew cannot expect to see or hear ivorybills, even if they are present, but they should be able to interpret the sights and sounds of a southern forest both to avoid false detections and to correctly identify an ivorybill if the opportunity arises. The best searchers will have years of experience observing flying ducks, crows, hawks, owls, and Pileated Woodpeckers, so that these birds are not mistaken for ivorybills when they are glimpsed in the forest.

We are fortunate that Ivory-billed Woodpeckers leave clear signatures of their presence in the form of large cavities and feeding marks on trees. The preliminary search of an area suspected of having ivorybills is best done by moving over as much acreage as possible in a kayak or on foot, assessing the age and composition of the forest, and looking for feeding trees and cavities. If suspicious cavities are observed, then the size of cavity entrances should be estimated. One means to do this would be to carry a plastic rod marked with black and white bands at exactly 1-centimeter intervals. When an interesting cavity is found, this banded rod can be lashed to a sapling and hoisted as close to the cavity as possible. From a high-resolution digital image showing both the cavity and the banded rod, one could make an estimate of the size of the cavity. If the banded rod is far below the cavity in the photo, such a digital image would not allow a precise measurement of cavity entrance size. But even if the measurement of a cavity entrance is imprecise, it should allow a searcher to decide if the cavity is in the 5-inch, 4-inch, or 3-inch size range. If measurements from digital images suggest that there are large cavities in the area, then more careful measurements of cavity sizes could be made on subsequent trips.

In my opinion, the best evidence for ivorybills in an area is the scaling of tightly adhering bark. At present, our interpretation of bark scaling in our ivorybill site is based on supposition and on comparison of the bark scaling in other southern bottomland forests where ivorybills don't occur. We need to more directly link scaled bark to foraging ivorybills. If such links can be made, then searching for bark scaling will certainly be the most efficient means to search bottomland forests for ivorybills. Searching for bark scaling was the primary technique that Tanner used when he searched southern forests for ivorybills in the 1930s. If suspicious bark scaling is found in an area being searched, then a sample of trees from the area should be measured for bark adhesion. If strong bark adhesion is measured on several scaled trees, and especially if such scaled trees occur in association with large cavities, then this suggests that the site has ivorybills, and a more concerted search of the area, perhaps with the deployment of listening stations, would be warranted.

Where should the next searches be focused? The entire Choctawhatchee River basin from above Geneva, Alabama to Choctawhatchee Bay needs to be carefully searched. Beyond the Choctawhatchee River basin, I think there is a good chance that ivorybills persist along several river basins in the Florida Panhandle. I am anxious to explore all of the river systems between Mobile Bay and the Apalachicola River. The most likely area to search next is the Apalachicola/Chipola River basin, which is only about 50–60 miles east of the Choctawhatchee River. Timothy Spahr and some colleagues spent a couple of weeks searching the Apalachicola River for ivorybills in 2003 and wrote a report on this search effort in *North American Birds*. Spahr reported extensive old-growth forest along portions of the Apalachicola and its main tributary the Chipola River, and they heard possible double knocks and kent calls along the Apalachicola. I am anxious to follow up on this expedition to see if we can confirm the presence of ivorybills in Apalachicola River bottomlands. If a population of ivorybills does persist along the Apalachicola River, then we can begin to be optimistic about the future of this species. If we had two populations of ivorybills in adjacent river systems, we could create habitat corridors between the drainages. In fact, swamp forest along Holmes Creek appears to already span much of the gap between the Choctawhatchee and Apalachicola Rivers. Ivory-billed Woodpeckers in multiple adjacent drainages would be an ideal situation for the maintenance of a viable population.

The Apalachicola River has long been a focus of ivorybill hunters. What about other river systems that, like the Choctawhatchee River, might have been overlooked? There are several likely candidates. The most likely overlooked river system that has tremendous ivorybill potential is the Escambia and its major tributary, the Conecuh River. As I mentioned earlier in the book, the only ivorybill specimen collected between the Alabama River drainage and the Apalachicola River was shot along the Conecuh River in Alabama in 1907. The Escambia River, into which the Conecuh River flows, appears on maps and on satellite images to be the twin of the Choctawhatchee River. It is an undammed river with a wide border of swamp forest and many oxbows and side channels. Like the forests along the Choctawhatchee, a large portion of the forests along the Escambia River have been protected by the Northwest Florida Water Management District for decades. To my knowledge, the Escambia River has never been searched for ivorybills.

The Yellow River and its major tributary, the Shoal River, have forested floodplains that are not as broad as those of the Choctawhatchee and Escambia rivers, but nonetheless these rivers are bordered by swamp forest as they traverse the Florida Panhandle. More intriguing, the Yellow and Shoal rivers border Eglin Air Force Base and the largest stands of old-growth longleaf pine in the world. If ivorybills use old-growth longleaf as foraging habitat, as some of the old literature suggests, then the Yellow and Shoal rivers might provide excellent habitat. A sight record of a pair of ivorybills from an area adjacent to Eglin Air Force Base near the Yellow River by Bedford Brown and Jeffrey Sanders in 1966 only adds to the allure of this area as a place to look for ivorybills.

The 1966 sighting is very intriguing, especially now, but the more I learn about this record, the less I'm willing to believe it. According to Ken Able, who was a prominent birder as well as ornithology professor at the time, Brown and Sanders were active birders in the Chicago area in the 1960s and 1970s. They nearly always birded together, and over a period of several years they reported a number of very rare birds around the Chicago area. Ultimately, these reports were shown to be fraudulent, Brown and Sanders confessed, and a long retraction of records appeared in the pages of *Audubon Field Notes*, the forerunner of *North American Birds*. The 1966 ivorybill sighting along the Yellow River occurred during the time of the incidents in Chicago.

The idea that there might be a self-sustaining population of Ivory-billed Woodpeckers in the Florida Panhandle makes me optimistic about the future of this species. Over the next couple of years, my students and I will be censusing all of these areas for ivorybills. Let's hope that other populations of ivorybills were overlooked across the Florida Panhandle.

Summary of sightings and sound detections made along the Choctawhatchee River, May 2005–May 2006

Summary of observations of Ivory-billed Woodpeckers near the Choctawhatchee River

Observer	Date	Time	Distance to bird	Circumstances	Traits observed
BR	5/21/05	07:30	50 m	Naked eye; fleeing bird	Large black woodpecker with white trailing edge on upper and under wing
TH	5/27/05	09:00	70 m	Binoculars; side view, then clear dorsal view	Loonlike flight; stiff wingbeats; white trailing edge on underwing; white trailing edge on upper wing; white lines running from neck to flank on each side dorsal surface; head (crest), back, rump, tail, and neck black
BR	7/31/05	10:30	15 m	Naked eye; overhead silhouette	Large woodpecker; stiff wingbeats; long neck/bill, long tail, long wings; no color
BR	7/31/05	11:00	15 m	Naked eye; overhead silhouette	Two large woodpeckers, one trailing other by 1 or 2 sec; on both: long wings; long neck/bill, long tail, black and white on wings
TH	12/24/05	15:30	50 m	Naked eye; head-on view, then banking left	Ducklike flight; underside of wings with white trailing edge and wing lining; black line running down center of underwing spreading to cover primaries

(continued)

Observer	Date	Time	Distance to bird	Circumstances	Traits observed
BR	12/27/05	10:37	30–40 m	Naked eye; fleeing bird	White secondaries, large, black woodpecker
GH	1/5/06	06:38	15 m	Naked eye; overhead silhouette	Fast, straight, loonlike flight; big woodpecker with long relatively narrow wings, long head/neck, long tail
GH	1/21/06	09:55	12–130 m	Naked eye; flying directly away	Powerful straight-line flight; from 2 m above water to 30 m in steady rise; large, black woodpecker with bright white trailing edge on dorsal wing; double knock heard
BR	1/25/06	14:54	100 m	Binoculars; fleeing bird	White secondaries, large black woodpecker; 2 double knocks heard
BR	2/1/06	08:51	60 m	Naked eye and binoculars (blurry); 2 birds just spooked off tree, clear underside view and side view	Two large woodpeckers coming off tree; clear view of underside of both birds wings; long, white wings with black line extending down center and widening toward wing tip; deep wingbeats as they hovered; shallow beats with direct flight once moving, flashing white; long necks
BR	2/2/06	16:12	50 m	Binoculars; low from tree, fleeing	Low, fast flight; white secondaries; large; long, slender wings and long neck
BR	2/26/06	14:52	100 m	Naked eye; far away, fleeing	Large woodpecker; white secondaries
BR, KS	4/27/06	16:30	60–80 m	Naked eye; flying across creek in front of canoe, poor lighting	Shape only; large bird, long wings, fast direct flight, ducklike; shallow beats
BR	5/19/06	16:30	10–15 m	Naked eye; flying across channel and swooping up to land on trunk of tree but not seen on tree	Large, black woodpecker with broad white trailing edge to wing; landing on side of tree trunk as a woodpecker

Legend: BR, Brian Rolek; TH, Tyler Hicks; GH, Geoff Hill; KS, Kyle Swiston.

Summary of sound detections of Ivory-billed Woodpeckers
near the Choctawhatchee River

Observer	Date	Time	What was heard
GH	5/21/05	08:45	double knock
TH, BR	5/22/05	09:00	kent call
Leon Hicks	12/21/05	09:00	kent call
TH	12/22/05	08:15	5 kent calls
TH, BR	12/23/05	10:30	double knock
TH	12/24/05	12:00	double knock
TH, BR	12/25/05	07:30	9 double knocks
TH, BR	12/26/05	16:30	double knock
Chet Gresham	1/6/06	09:10	double knock
TH	1/13/06	16:30	2 kent calls
BR	1/13/06	16:30	kent call
BR	1/20/06	09:08	15–20 double knocks
BR	1/20/06	11:30	1 double knock
GH	1/21/06	09:57	double knock
BR	1/23/06	01:30	6 double knocks
BR	1/25/06	14:54	2 double knocks
KS	1/27/06	10:45	1 double knock
GH	2/4/06	14:55	4 double knocks
BR	2/7/06	12:07	12 kent calls
KS	2/8/06	07:10–07:35	50 kent calls
BR, KS	2/8/06	11:15	20 kent calls
BR	2/19/06	15:45–16:15	45 double knocks
GH	2/24/06	17:00	4 kent calls
GH	3/3/06	08:30	12 kent calls
GH, David Carr	3/3/06	17:22	kent call, double knock
David Carr	3/4/06	06:34	double knock
BR	3/8/06	06:30–07:15	8–10 kent calls
KS	3/8/06	06:30–07:15	5–8 kent calls
BR	3/9/06	17:19	5 kent calls

(continued)

Observer	Date	Time	What was heard
BR	3/12/06	15:30	2 kents
BR	3/17/06	14:52	2 double knocks
TH, BR	3/19/06	15:55	10 double knocks (TL heard 4)
TH	3/19/06	16:33	10 double knocks
BR	3/20/06	12:30	3 double knocks
TH	3/21/06	08:30	kents
TH, BR	3/24/06	15:45	4 kents (TH heard 1 double knock)
KS	3/29/06	08:30	10 double knocks
BR	4/1/06	17:10	1 double knock
BR	4/7/06	14:55	5 double knocks
BR	4/14/06	10:00	1 double knock
BR	4/24/06	17:00	2 double knocks

Legend: BR, Brian Rolek; TH, Tyler Hicks; GH, Geoff Hill; KS, Kyle Swiston.

SUGGESTED READING

In this book, I give a narrative of a search for Ivory-billed Woodpeckers. I wanted this to be a readable account that could be enjoyed by nonacademic bird enthusiasts, so I did not cite the scientific literature. Below I suggest some books and web pages on Ivory-billed Woodpeckers to which interested readers might turn. For anyone interested in scientific papers on Ivory-billed Woodpeckers, the references in Jackson (2006) are comprehensive.

Tanner, J. T. 1942. The Ivory-billed Woodpecker. Research Report Number 1. National Audubon Society, New York.

> [*The only detailed account of the natural history of the Ivory-billed Woodpecker based on a study in the Singer Tract in Louisiana. Also includes a summation of Ivory-billed Woodpecker populations in 1939.*]

Hoose, P. 2004. The Race to Save the Lord God Bird. Farrar, Straus and Giroux, New York.

> [*An entertaining account of various human interactions with Ivory-billed Woodpeckers both in North America and Cuba.*]

Gallagher, T. 2005. The Grail Bird: Hot on the Trail of the Ivory-billed Woodpecker. Houghton Mifflin, Boston.

> [*Primarily a summary of ivorybill sightings in the Mississippi River Valley after the cutting of the Singer Tract. The author moves from investigative*

journalist to the center of action when he and a colleague actually spot an ivorybill, setting off one of the greatest searches for a wild animal in modern history.]

Jackson, J. A. 2006. In Search of the Ivory-billled Woodpecker. Harper-Collins Publishers, New York.

[*The most thorough academic treatment of the natural history and conservation history of Ivory-billed Woodpeckers.*]

Web pages of interest

http://www.auburn.edu/ivorybill

[*Describes the search for Ivory-billed Woodpeckers in the Choctawhatchee River basin by my lab group at Auburn University.*]

http://web2.uwindsor.ca/courses/biology/dmennill/ibwo

[*Describes the sound monitoring being conducted in the Choctawhatchee River basin by Dan Mennill and his students at the University of Windsor.*]

http://ibwo.blogspot.com/

[*Blog of the coordinator of the volunteer team searching for ivorybills along the Choctawhatchee River in 2007.*]

http://www.birds.cornell.edu/ivory/

[*Not only gives accounts of the search for ivorybills in the Big Woods of Arkansas, but has lots of general information about ivorybills, including historic film and sound clips.*]

INDEX